Heat Transfer 2

Mathematical and Mechanical Engineering Set

coordinated by
Abdelkhalak El Hami

Volume 10

Heat Transfer 2

Radiative Transfer

Michel Ledoux
Abdelkhalak El Hami

iSTE

WILEY

First published 2021 in Great Britain and the United States by ISTE Ltd and John Wiley & Sons, Inc.

ISTE Ltd
27-37 St George's Road
London SW19 4EU
UK

www.iste.co.uk

John Wiley & Sons, Inc.
111 River Street
Hoboken, NJ 07030
USA

www.wiley.com

Library of Congress Control Number: 2020949611

British Library Cataloguing-in-Publication Data
A CIP record for this book is available from the British Library
ISBN 978-1-78630-517-6

Contents

Preface

Thermal science is to thermodynamics as decree is to law. It answers the following question – which all good leaders must (or should) ask themselves whenever they have an "idea": "How would this work in practice?". In a way, thermal science "implements" thermodynamics, of which it is a branch. A thermodynamics specialist is a kind of energy economist. Applying the first principle, they create a "grocery store". With the second principle, they talk about the quality of their products. I add or remove heat from a source or work from a system. And the temperature, among other things, defines the quality of the energy for me.

But *by what means do I take or do I give*? Even calculations of elementary reversible transformations do not tell us by what process heat passes from a source to a system.

Thermal science specifies how, but "evacuates" work. If in a given problem related to, for example, a convector where an electrical energy (therefore in the "work" category) appears, it is immediately dissipated into heat by the Joule effect.

Three heat transfer modes can be identified: conduction and radiation – which can be seen separately, although they are often paired up – and convection, which is by nature an interaction of fluid mechanics and conduction.

Dividing the study of thermal science into three is the result of logic. Presenting this work in three volumes is somewhat arbitrary; in our opinion, however, this split was necessary in order to keep the volumes in the collection a reasonable size.

This is Volume 2 of a collection of problems on thermal transfer, dedicated to radiation and digital approaches to transfer. Even though it is primarily a collection of exercises, a great deal of attention is focused on lessons. For the most part, the work is a first introduction to the thermal calculation of practical devices, which may be enough in itself. For subsequent calculations, the reader will still have to turn to specialist works or encyclopedias available in the field of thermics.

In **Chapter 1**, after a brief historical background, we summarize the vital notions of electromagnetic radiation and how they are written. The emphasis in this book is on the aspect of energy: the notions of photometry prove indispensable at this stage of exploration.

At the heart of studying radiation, **Chapter 2** focuses on calculating luminances, relying on black body laws (Planck's law; Rayleigh–Jeans and Wien approximations) and its derivatives: Stefan–Boltzmann laws and Wien laws. For evaluating a fraction of total emittance radiated in a spectral band, the G_{λ_1,λ_2} function proves vital.

Chapter 3 tackles these interactions between a light flux and a material medium, a fundamental subject in any practical calculation of radiation: the phenomena of emission, absorption, transmission, etc., as well as the Kirchhoff law, emissivity, absorption coefficient, etc.

Chapter 4 presents the general notions on reciprocal radiation from several surfaces. We distinguish total influence and reciprocal radiation from finite surfaces. This subject is central in particular to the calculation of ovens. Here, we should restrict ourselves to notions, returning to specialist works for application by professionals working with ovens.

Solving a radiation problem often involves knowing and understanding the essence of the corresponding lessons. It is therefore not so easy to produce (interesting) problems that are limited to only being a paragraph long. This is why we were led to focus in **Chapter 5** on the essence of exercises dedicated to radiation and coupled transfers.

In a domain where digital methods are becoming the rule for complex situations, it seems important to reserve a particular place for “monitoring” analytical approaches. In addition to their usefulness in understanding this domain, they offer the reader a precious tool for making calculations “on the back of an envelope”.

Finally, **Chapter 6** introduces the reader to a digital approach for different transfer modes. The general problem of modeling is tackled here and examples of processing using ANSYS are presented.

The **Appendix** covers the tabulations of $G_{O-\lambda T}$ functions whose practical importance emerges from the problems.

December 2020

Introduction

I.1. Preamble

Thermal energy was probably first perceived (if not identified) by humanity, through the Sun. The themes of night and day are found at the center of most ancient myths. Humanity's greatest fear was probably that the Sun would not return again in the morning. Fire became controlled in approximately 400,000 BP. Thermal transfer was therefore a companion of *Homo ergaster*, long before *Homo sapiens sapiens*.

However, it took a few hundred thousand years before so-called "modern" science was born. Newtonian mechanics dates from three centuries ago. Paradoxically, another century and a half passed by before energy was correctly perceived by scientists, in terms of the new field of thermodynamics. Furthermore, a systematic study of heat transfer mechanisms was carried out at the end of the 19th century, and even later for the study of limit layers, the basis of convection.

Heating, lighting and operating the steam engines of the 19th century were all very prosaic concerns. Yet this is where revolutions in the history of physics began: the explosion of statistical thermodynamics driven by Boltzmann's genius, and quantum mechanics erupted with Planck, again with Boltzmann's invovlement.

Advances in radiation science, particularly in sensor technology, has enabled us to push back our "vision" of the universe by a considerable number of light years. To these advances we owe, in particular, the renewed interest in general relativity that quantum mechanics had slightly eclipsed, through demonstration of *black holes*, the physics of which may still hold further surprises for us.

Closer to home, fundamental thermal science, whether it is conduction, convection or radiation, contributes to the improvement of our daily lives. This is particularly true in the field of housing where it contributes, under pressure from environmental questions, to the evolution of new concepts such as the active house.

The physics that we describe in this way, and to which we will perhaps introduce some readers, is therefore related both to the pinnacles of knowledge and the banality of our daily lives. Modestly, we will place our ambition in this latter area.

There are numerous heat transfer textbooks in different formats: "handbooks" attempting to be exhaustive are an irreplaceable collection of correlations. High-level courses, at universities or engineering schools, are also quite exhaustive, but they remain demanding for the listener or the reader. Specialist, more empirical thematic manuals are still focused on specialists in spite of all this.

So why do we need another book?

The authors have taught at university level and in prestigious French engineering schools, and have been involved in the training of engineers on block-release courses. This last method of teaching, which has been gaining popularity in recent years, particularly in Europe, incorporates a distinctive feature from an educational point of view. Its practice has, in part, inspired this book.

The aim is to help learners who have not had high-level mathematical training in their first years following the French Baccalaureate (therefore accessible to apprentices), and pupils with more traditional profiles. At the same time, we would like to show this broad audience the very new possibilities in the field of digital processing of complex problems.

When a miner wants to detach a block of coal or precious mineral from a wall, they pick up a pneumatic drill. If we want to construct a tunnel, we must use dynamite. The same is true for physicists.

Whether they are researchers, engineers or simply teachers, scientists have two tools in their hands: a *calculator* and a *computer* (with very variable power). Since both authors are teacher researchers, they know they owe everything to the invention of the computer. From the point of view of teaching, however, each one of the two authors has remained specialist, one holding out for the calculator and "back-of-the-napkin" calculations, and the other one using digital calculations.

The revolution that digital tools has generated in the world of "science" and "technical" fields, aside from the context of our daily lives, no longer needs to be proved. We are a "has been" nowadays if we do not talk about Industry 4.0. The "digital divide" is bigger than the social divide, unless it is part of it....

Indeed, the memory of this revolution is now fading. Have students today ever had a "slide rule" in their hands? Do they even know what it is? Yet, all the physicists behind the laws of thermal science had only this tool in hand, giving three significant figures (four with good visibility and tenacity), leaving the user to find the power of ten of the result. It goes without saying that a simple calculation of a reversible adiabatic expansion became an ordeal, which played a part in degrading the already negative image of thermodynamics held by the average student.

This reminder will seem useless to some; slide rules are at best sleeping in drawers. But there is a moral to this story: no matter what type of keyboard we type on, a calculator or a computer, our head must have control over our fingers. This book has been written on the basis of this moral.

A good physicist must have a perfect understanding of the idea of an "order of magnitude". For this, the tool is a calculator. We always do a rough sizing of a project before moving on to detailed modeling and numerical calculations.

The two authors belong to the world of engineering sciences, meaning most of their PhD students have entered the private sector. One of them, having moved into the aerospace sector, came back to see us very surprised by the recurrence of "back-of-the-napkin" calculations in his day-to-day work.

Fundamental or "basic physics" concepts are taken from a type of manual that is resolutely different from those dedicated to the numerical approach. In this case, the authors allow themselves to believe that it is no bad thing to collect them all together in a single book, for once. This is a significant difference that will surprise some and, without doubt, be criticized by others. Nevertheless, when reading this book, an "average" student will be initiated to a field that teaching models generally promised "for later on" (or never if he/she never goes beyond a certain level of education). It is also true that fully immersed in equations and complex calculations, specialist readers will be able to "be refreshed" when faced with the short exercises, which can sometimes surprise and encourage them – why not – to go back to their roots (assuming they had indeed been there).

Another significant difference is that this book is directed at a large scientific audience, which covers possibly the entire field: researchers, PhD students or those who have obtained Confirmation and are just starting out in the field, technicians, students or professionals, engineers. This last type of scientist is perhaps the main target of this book.

So, what is this book for?

Above all, it **contains problems to be worked on**, of which most are accessible to all, from the level of an apprentice technician upwards, either one or two years after the Baccalaureate. This book was written in France, where scientific teaching is structured around universities, engineering Grandes Écoles, engineering training through apprenticeships and two types of technician training sections at high schools or universities. In countries with simpler models, readers should also find it useful.

It seems necessary to surround these problems with strong reminders of past learning, so that the reader does not need to permanently refer back to their manuals. We see two advantages in this: a presentation of the scientific material focusing on the problems, and a second chance for readers to integrate notions that perhaps had not been well understood in the initial teaching.

Lastly, upon rereading, the authors also recommend this book as an introduction to the taught disciplines.

I.2. Introduction

Thermal science is to thermodynamics as decree means is to law. It answers the following question – which all good leaders must (or should) ask themselves whenever they have an “idea”: “How would this work in practice?”.

In a way, thermal science “implements” thermodynamics, of which it is a branch.

A thermodynamics specialist is a kind of energy economist. Applying the first principle, they create a “grocery store”. With the second principle, they talk about the quality of their products. I add or remove heat from a source or work from a system. And the temperature, among other things, defines the quality of the energy for me.

But *by what means do I take or do I give*? Even calculations of elementary reversible transformations do not tell us by what process heat passes from a source to a system.

Thermal science specifies how, but “evacuates” the work. If in a given problem related to, for example, a convector where electrical energy (therefore in the “work” category) appears, it is immediately dissipated into heat by the Joule effect.

Three heat transfer modes can be identified: conduction and radiation – which can be seen separately, although they are often paired up – and convection, which is by nature an interaction of fluid mechanics and conduction.

Dividing the study of thermal science into three volumes is the result of logic. Presenting this work in three volumes is somewhat arbitrary; in our opinion, however, this split was necessary in order to keep the volumes in the collection a reasonable size.

The first volume, entitled **Heat Transfer 1**, is dedicated to “classic” approaches (analytical treatment) to conduction, which will be of greater interest to readers who are looking for “simple” prediction methods.

The second volume, entitled **Heat Transfer 2**, is dedicated to “classic” approaches (analytical treatment) of radiation, and assembles digital approaches of these various transfer modes. It is aimed at engineers or researchers who want to resolve more complex problems.

The third volume, entitled **Heat Transfer 3**, is focused on convection transfers. As we have already pointed out, all of these transport operations are rarely pure and lead to problems that involve three inter-connecting transfer modes, conduction, convection and radiation.

Before our readers immerse themselves in a text that, despite our best efforts, remains intellectually demanding, we propose a short text that is a little lighter.

I.3. Interlude

Let us imagine, in a "B movie" context, a somber hostel in the gray fog of a port in the middle of nowhere. Sailors from a faraway marina come and drink away their troubles. And as always, the drink helping them along, they turn to fighting.

Let us entrust Ludwig Boltzmann to direct the film. Our B movie heroes are getting agitated, delivering blows to one another. Each one of them has moderate kinetic energy, distributed heterogeneously among them in the room. For some reason, they get involved in a general brawl. Their average kinetic energy becomes much greater.

In everyday language, we would say that **things are hotting up**.

This would bring us right into line with a fundamental concept of Boltzmann, who was the first to hypothesize that **heat is made up of molecular agitation**. *The temperature in a gas is proportional to the average quadratic energy of the molecules that make it up*:

$$E_C = \frac{1}{2} k T$$

Using this model, we will return to the physical basis for all transport phenomena.

On the way, we rarely escape from the explosion of a door or a window, giving in under the repeated beatings of the brawlers.

We have just modeled the pressure, due to the transfer of the quantity of movement on the surface, by the impact of molecules.

Let us now imagine that the altercation is initially located in the corner of the room: a smaller group starts fighting between themselves.

From kicks to punches, after multiple impacts within the group and its immediate neighbors, the agitation will spread: *we have just seen the mechanism of heat propagation by transfer of impacts.*

Let us place an imaginary separation (geometrically but immaterially defined) at the center of the room. Let us count the sailors that cross through it within a unit of time.

This wall is now crossed by kinetic energy: we have defined a flow of heat.

Let us put a metal ring with a surface area of $S = 1\,m^2$ in the room. On both sides of this ring, the blows exchanged constitute a transfer of kinetic energy – we have just defined **the heat flow density**.

And we have just understood the nature of the propagation of thermal flows by impacts.

Let us suppose that the great majority of the brawlers come from a ship with a white uniform. Let us suppose that another boat in the port has uniforms that are red. The red ones are initially all united. We will then quickly see that the red mariners, as they receive and return blows, spread out across the room. *We have just shown the mechanism of* **diffusion of matter**, *of a component within a mixture.*

We will have a better qualitative understanding that the fundamental law of conduction (Fourier Law) is formally identical to the law for the diffusion of mass (Fick Law).

Let us put our agitated sailors in the compartments of a flatcar train, where they continue to fight. And let us start the train moving. The kinetic energy that they contain is transported from one point to another.

We have just invented **thermal convection**.

We can go further.

Let us imagine a series of flatcar trains on a set of parallel tracks. The train furthest to the side is fixed to a platform. All of these trains are full of sailors. Let us suppose that our train follows the outside, parallel rail tracks. No brakes will prevent these trains from moving. Only the last train, at the platform, is stuck.

For a reason we do not need to analyze (cinema allows all kinds of fantasy), "clusters" of fighting sailors jump from one wagon to the next. These "clusters" contain a component of speed that is parallel to the train, which will communicate information about the quantity of movement to the adjacent train. These trains will then start to move, more quickly the closer they are to the outside train. And the same occurs up to the train at the platform. This train will not move, but a force will be applied to its brakes.

We have just discovered the mechanism of dynamic viscosity. At the same time, the parallel trains in relative movement give us a picture of the *notion of boundary layers.*

At the same time, these agitated clusters carry their disordered kinetic energy, "thermal" agitation. We have just seen the mechanism of the *thermal boundary layer.*

Finally, let us include a few red mariners in the crowd of white. They will be carried with the clusters, and we have just invented *the limit layer of diffusion of a species.*

We are in a fantasy, and let us benefit from it as far as we can. To finish, let us suppose that this is carnival day; each sailor has a belt equipped with bells.

All the individuals have a different speed, and the impacts are random, all the bells start to jingle, each with a different frequency. The distribution of frequencies will depend on the statistical distribution of speeds (Boltzmann statistics), and the intensity of noise produced will depend on the total agitation energy of the sailors.

We have just understood the **basic mechanism of radiation**. *We have just realized why the theory of radiation needed to use the concepts of* *statistics derived from the work of Boltzmann – a brilliant pupil of Planck – to produce the emissions spectrum of a black body, for example.*

NOTE.– *The model is certainly simplistic. The emission comes from quantum transitions in the gas atoms.*

Here, we have already deviated from the pure substance of the book, but we could go even further.

Let us suppose that our agitated sailors are in a room with one mobile wall (a nightmare scenario frequently seen on the silver screen).

The incessant impacts of the fighters on this wall create a force that pushes it. This force, reduced to a surface unit, explains the notion of pressure.

By pushing against this wall, our crowd applies work that is greater than the resistance.

Here we see an equivalence spring up between work and heat that, at a fundamental level, are simply two mechanical energies: one ordered and the other disordered. **The first principle of thermodynamics is illustrated by this.**

We can see that the incidence of an average blow on the wall is rarely normal.

Therefore, an average fighter will have a trajectory that will be reflected off the wall. And only the normal component of its speed will be able to push (or transfer work to) the wall.

Thus, we see that it will be impossible for the crowd (taken to mean a gas) to give all its energy to a mobile wall.

The fundamental mechanism that leads to the **second principle of thermodynamics** *has just been demonstrated.*

These "light-hearted" images, which will perhaps not please everyone, were an oral support for the presentation of different transport phenomena by one of the authors. We hope that the reader, once they have studied this book, will want to return to this text. They will then have understood, we hope, the images that lead to the development of thermodynamics.

And if this text has a moral, it would be: *Writing down thermodynamics, just like thermal science, is based on continuous equations. The fundamentals of physics that determine these phenomena arise from the field of the discontinuous: discontinuity of matter, divided into particles; discontinuity of light, divided into photons.*

1

General Remarks

1.1. Introduction

Any matter that is not at absolute zero emits radiation. This radiation is electromagnetic by nature. Electromagnetic radiation is characterized by its frequency or its wavelength as well as by the sizes of the electric and magnetic fields that it transports, or better, by the energy it transports. The entire wavelength or frequency spectrum can be used by radiation. Electromagnetic radiation has been modeled by Maxwell and thermal emission by Planck. We note that what we normally call light, more precisely visible light forms, is only a small part of a wavelength spectrum ranging from zero to infinity. Our concern here will be to decipher the radiative energy transmitted or received by material bodies, by linking it with radiations that can be identified by their wavelengths (or their frequencies). We will go over the definition of these terms. Moreover, the propagation of radiative energy involves a particular metrology called photometry. We should therefore introduce some common definitions from this domain.

1.2. Propagation of a sinusoidal electromagnetic wave

1.2.1. *Frequencies and wavelengths*

We will refer to some specialist work for a general study of wave propagations, as well as for the modeling of electromagnetic waves.

We will simply recall here the definitions of two importance parameters, the frequency ν and wavelength λ of a ray.

An electromagnetic radiation is manifested by the evolution in space of an electric field $E(x,t)$ and a magnetic field $B(x,t)$ paired, and so variable in time and space.

In the simplest case of a plane wave propagating along a rectilineal axis $0x$ (which results in the "light ray"), the space state is identical on any plane of abscissa x normal to this axis. The fields $E(x,t)$ and $B(x,t)$ are therefore expressed in the form of a function such that

$$f(x,t) = A\cos 2\pi\left(\nu t - \frac{x}{c}\right) = A\cos 2\pi\left(\frac{t}{T} - \frac{x}{c}\right) \quad [1.1]$$

This formula, which can be notably obtained by solving the laws of electromagnetism (Maxwell's theory), is thus interpreted:

At a given point, f is a wave function, of amplitude A. ν is the frequency of the perturbation observed, and T is the time for a movement or period.

We can "follow" the propagation in space. We seek the evolution in time of the abscissa of a given state (particular value of f). f is therefore a function of time and space, $f(\vec{r},t)$.

This perturbation moves at celerity c (celerity since it does not involve moving matter, but a state of perturbation). During a period T, the wave moves from a distance $\lambda = c\,T$, called a wavelength.

For electromagnetic waves, this celerity varies depending on the medium crossed. It has the form $\frac{c}{n}$, where n is the medium's refraction index. The celerity of the "light", often denoted as c, is a universal constant $c = 299\,792\,458\,m.s^{-1} \approx 300\,000\,km.s^{-1}$. This is the maximum possible celerity for an electromagnetic wave.

Moreover, in a void, c is the same for all wavelengths. For material bodies, the refraction index varies with the wavelength (dispersion phenomenon).

The wavelength of a radiation therefore depends on the medium traversed, unlike the frequency.

A radiation can be formed by several frequencies. When there is only one, we speak of a monochromatic radiation.

We can therefore characterize a monochromatic radiation in two ways:

– by its frequency ν;

– by its wavelength λ in the void.

Instinctively, we could consider that the frequency is the most logical characteristic of a radiation; this frequency is independent of the medium. The celerity of the light varies with the medium; it is equal to $V = \frac{c}{n}$, where n is the refraction index, which leads to the wavelength being redefined as $\lambda = \frac{c}{n\nu}$.

In the end, the wavelength, that is, the distance crossed during a period, is merely an indirect characteristic.

Nevertheless, considering orders of size here calls for a different choice.

We will consider a visible light, red for example, of $\lambda = 700\,nm = 7.10^{-7}\ m$. The radiation frequency is in the order of:

$$\nu = \frac{c}{\lambda}\lambda \sim \frac{3.10^{8}}{7.10^{-7}} = 4{,}3.10^{14}\ Hz \qquad [1.2]$$

We see that the orders of size of their wavelengths are far more comfortable to use than those of the frequencies.

In this book, we choose, like many other authors, to characterize radiations by their wavelengths λ in a vacuum. Note "in the vacuum", as in practice, we systematically omit this specification for simplicity.

1.2.2. *Radiation spectrum*

The quantity of energy carried by the wave is given by the knowledge of the As.

A general ray carries several waves of different wavelengths, each allocated an amplitude $A(\lambda)$. ***The data from $A(\lambda)$ constitutes a spectrum.***

We will see that in practice, this parameter $A(\lambda)$ will be expressed implicitly from the luminance $L_\lambda(T)$. The distribution of energy depending on the wavelength will define a spectrum.

This spectrum may be continuous or discontinuous. In a continuous spectrum, an energy density is attributed to each wavelength. A continuous spectrum can span the entire range of wavelengths, from $\lambda = 0\ m$ to $\lambda \to \infty$, where a body can only emit in one or more bounded area wavelengths or $\lambda = 700\ nm$ to $\lambda = 2000\ nm$, for example.

A body can only emit (this is the case with atoms) for some wavelengths, or rather, a limited number of small areas around a wavelength $\lambda_1, \lambda_2, \lambda_3, \ldots\lambda_n$, etc. We therefore speak of a radiation spectrum.

Electromagnetic waves have received denominations linked to their generating phenomenon, or to their use.

We can thus cite, following a range of decreasing wavelengths, that is, at increasing frequencies:

Radio frequency waves, from 3 mm to 33 km. Old radio stations at amplitude modulation were called by their wavelength. We distinguish long wave (LW), medium wave (MW) and short wave (SW) amateur radio domains expressed in wavelength. The dials for our receivers in "modern" radio, in frequency modulation, are graduated in MHz. We distinguish VHF (metrical waves, for example radio FM), UHF (centimetric waves, for example television) and SHF (millimetric waves, for example radar). We signal the case of specialist “microwaves” for heating (cooking, etc.), whose length is fixed at $\lambda = 12.2\,cm$ to avoid interference with radio.

The infrared domain from $700\,\text{nm}$ to $1\,\text{mm}$. We distinguish (ISO cutting, but there are others) near infrared ($0.78\,\mu\text{m} - 3\,\mu\text{m}$), mid infrared ($3\,\mu\text{m} - 50\,\mu\text{m}$) and far infrared ($50\,\mu\text{m} - 5\,\text{mm}$).

The visible light domain. The International Commission on Illumination defines the vision of the reference observer for a wavelength in a vacuum of $380\,\text{nm}$ to $780\,\text{nm}$.

We can see from this domain, defined from the performances of a particular sensor, the human eye is very narrow.

The ultraviolet domain, comprised between the visible and X-rays. Often called black (since it is not visible) light, the Sun is its main source. They are responsible, among others, for tanning, but also for sunburn and cancers. The atmospheric ozone layer partially protects us from it.

We distinguish:

UV A ($400\,\text{nm} - 315\,\text{nm}$);

UV B ($315\,\text{nm} - 280\,\text{nm}$);

UV C ($280\,\text{nm} - 100\,\text{nm}$).

NOTE.– Although its wavelengths are higher than the visible spectrum, we use the term **infra**-red for radiations whose frequency is lower than that of visible light. In the same way, although its wavelengths are less than those of the visible spectrum, we use the term **ultra**-violet for radiations whose frequency is higher than visible frequency.

X-rays

They are generated mainly by the interaction of an electron beam and a metal plate (anti cathode). They are a vital element in medical imaging, and remain as so, even though other methods have emerged. Wavelengths vary from (10^{-8} m) to 10^{-12} m.

Gamma rays

Gamma rays of wavelengths lower than a picometer ($1\ \text{pm} = 10^{-12}\ \text{m}$).

NOTE.– The lowest wavelengths correspond to the highest frequencies, so do the photons with the most energy. This explains why for gamma ultraviolet rays, the harmful effects of the rays on human biology, in particular, increase.

We will see that in thermics, for temperature ranges encountered in practice, ray emissions involve visible and near infrared.

1.3. The concept of photometry

1.3.1. *Geometric parameters*

The light ray is a model used in geometrical optics.

From the perspective of optical physics, it represents a "carrier" of the wave vector $\vec{k}$ of norm:

$$k = \frac{2\pi}{\lambda} \qquad [1.3]$$

From the perspective of photometrics, light energy does not propagate along a line, but is contained in a volume. This volume is "bounded" by a cone formed of "rays". This cone of light is measured in terms of a solid angle.

Consider the intersection of a cone issuing from a source point with a unit radius sphere. Defined in this way from a source point, the solid angle Ω is the area of the fraction of the unit radius determined by the cone.

To show that the extent of the cone is being discussed, we do not use Ω in m^2 but in **steradians** *sr*.

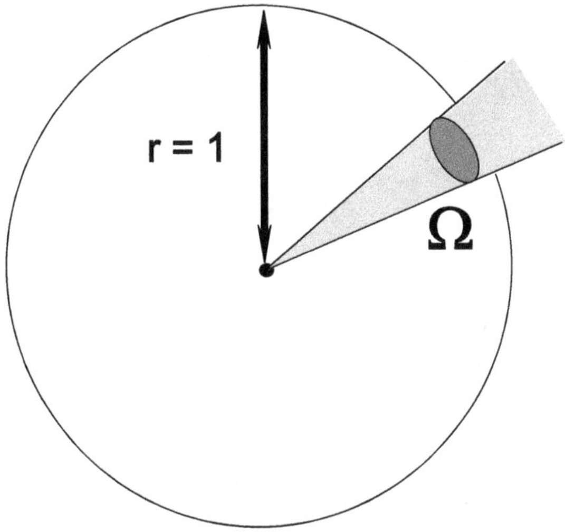

Figure 1.1. *Concept of the solid angle. For a color version of this figure, see www.iste.co.uk/ledoux/heat2.zip*

Information is often given on the solid angle from the area S, the intersection of the cone and a sphere of radius $R \neq 1$. The intersection of this cone with a sphere of radius $r = 1$ (with the area defining the solid angle, Ω) will be in homothety with S. For surfaces, this homothety will involve a relationship:

$$\frac{S}{\Omega} = \frac{R^2}{1} = R^2 \qquad [1.4]$$

For this reason, the small solid angle determined by a center O, a small spherical cap with area dS centered on O, will be:

$$d\Omega = \frac{dS}{R^2} \qquad [1.5]$$

This is particularly true for a cone with an infinitely small angle, which is determined from a surface dS centered in O' normal to the ray OO'. The solid angle is then expressed by:

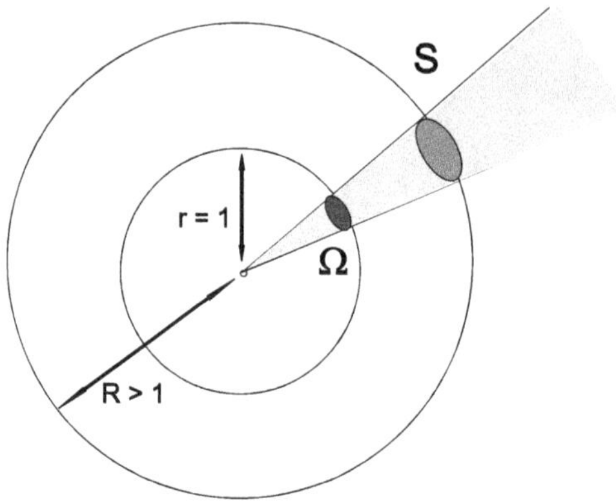

Figure 1.2. *Illustration of* $d\Omega = \frac{dS}{R^2}$. *For a color version of this figure, see www.iste.co.uk/ledoux/heat2.zip*

If the surface dS is not normal to OO', but has a normal that makes an angle θ with OO', the solid angle is calculated with the projection of dS on a normal plane at OO', which will then have an area of $dS\cos\theta$. The solid angle will then be expressed by:

$$d\Omega = \frac{dS\cos\theta}{R^2} \qquad [1.6]$$

Moreover, it shows that the solid angle defined by a half-angle cone at the top α is:

$$\Omega = 2\pi(1-\cos\alpha) \qquad [1.7]$$

which for a small angle gives:

$$d\Omega = 2\pi\sin\alpha\, d\alpha \qquad [1.8]$$

For the whole space, the solid angle becomes the surface of a sphere with radius $R=1$:

$$\Omega = 4\pi R^2 = 4\pi\ sr \qquad [1.9]$$

The half-space is therefore equal to $2\pi\ sr$.

1.3.2. *Radiance*

The energy radiated by a surface dS in a direction that makes an angle θ with the normal to this surface element varies with:

– the absolute temperature T of the emitting surface;

– the wavelength interval $d\lambda$, or spectral range, one in which we express the emission.

Moreover, we return the transmission to a temperature T in a spectral interval $d\lambda$ to a normal surface to the axis of the basic solid angle $d\Omega$ considered. This surface has an area of $dS\cos\theta$.

The energy sent by a surface dS in a solid angle $d\Omega$, whose axis makes an angle θ with the normal to dS, is expressed in the form of a flow $d^3\Phi$ that is written as:

$$d^3\Phi = L_{T,\lambda}(\theta)\, dS\cos\theta\, d\lambda\, d\Omega \qquad [1.10]$$

This flow is differential three times, through dS, $d\lambda$ and $d\Omega$.

$L_{T,\lambda}(\theta)$ is the monochromatic radiance of the surface in the direction θ. This radiance is defined by the preceding expression.

NOTE.– The expression $L_{T,\lambda}(\theta)$ is common in books on radiation. It is clearly mathematically equivalent:

$$L_{T,\lambda}(\theta) \equiv L(T,\lambda,\theta) \qquad [1.11]$$

This particular writing highlights the fact that, in all instances, radiance is variable with T and λ, while it is not in all cases with θ (this will be the particular, important case expressed by Lambert's law). It should be noted that the angle θ is involved twice, in two different ways in the formula $[1.10]$:

– the θ figuring in $L_{T,\lambda}(\theta)$ marks the influence of the ray angle with the surface on electromagnetic emission (encrypted by the luminance);

– the θ figuring in $d\,S\cos\theta$ does not mark a variation of the emission with θ, but provides a geometric reference to a normal surface at the axis of the elementary solid angle $d\,\Omega$ considered.

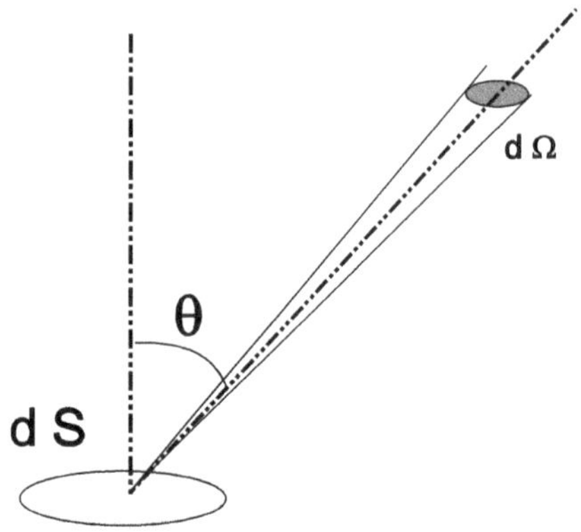

Figure 1.3. *The angle θ, For a color version of this figure, see www.iste.co.uk/ledoux/heat2.zip*

When we look at the energy emission of a surface over the whole spectrum of light, we integrate it over the wavelengths and have:

$$d^2\,\Phi = L_T\,(\theta)\,d\,S\cos\theta\;d\,\Omega \qquad [1.12]$$

with:

$$L_T\,(\theta) = \int_0^\infty L_{T,\lambda}\,(\theta)\,d\,\lambda \qquad [1.13]$$

This defines a total radiance.

We note that the flow $d^2\,\Phi$ in this instance is only twice differential.

NOTE.– Writing the radiance in $L_{T,\nu}\,(\theta)$

We do not use this writing here, but it can be found, especially in articles on radiation transfer. In this case, it is a good idea to know how to link them.

The flow expression $d^3\Phi$ can also be written in terms of a spectral frequency interval, *v*:

$$d^3\,\Phi = L_{T,\nu}\left(\theta\right) d\,S \cos\theta\, d\nu\, d\,\Omega \qquad [1.14]$$

The expression of $L_{T,\nu}\left(\theta\right)$ will then be different from that of $L_{T,\lambda}\left(\theta\right)$.

The relationship between these two radiance expressions is simple to find; noting that, in a set spectral interval, the energy mission should retain the same value, so:

$$\begin{aligned} d^3\,\Phi &= L_{T,\nu}\left(\theta\right) d\nu\, d\,S \cos\theta\; d\,\Omega = L_{T,\nu}\left(\theta\right) \\ d\nu &= L_{T,\lambda}\left(\theta\right)\, d\lambda\;\, d\,S \cos\theta\;\, d\,\Omega \end{aligned} \qquad [1.15]$$

We then identify:

$$L_{T,\nu}\left(\theta\right) d\nu \; = L_{T,\nu}\left(\theta\right)\; d\nu \; = L_{T,\lambda}\left(\theta\right)\; d\lambda \qquad [1.16]$$

Noting that:

$$\lambda = \frac{c}{\nu}\,, \qquad [1.17]$$

where c is the celerity of the light. We differentiate both terms of equality:

$$d\,\lambda = -\frac{c\,d\,\nu}{\nu^2} \qquad [1.18]$$

where

$$d\,\nu = -\frac{c\,d\,\nu}{\lambda^2} \qquad [1.19]$$

which is the relationship between $L_{T,\nu}\left(\theta\right)$ and $L_{T,\lambda}\left(\theta\right)$:

$$L_{T,\nu}\left(\theta\right)\; d\nu \; = -\frac{c\,d\,\nu}{\nu^2} L_{T,\lambda}\left(\theta\right) \qquad [1.20]$$

Or indeed

$$L_{T,\lambda}(\theta) = -\frac{c}{\lambda^2} L_{T,\nu}(\theta) \qquad [1.21]$$

1.3.3. *Bouguer–Lambert law*

We consider two infinitely small surfaces dS_1 and dS_2 facing one another, centered on two points O_1 and O_2. Their centers are distant by $r = O_1 O_2$; this parameter is large compared to area size scales. The normal to dS_1 makes an angle θ_1 with $O_1 O_2$. The normal to dS_2 makes an angle θ_2 with $O_1 O_2$.

dS_1 is the light-emitting surface, and dS_2 is the receiving surface. What is the flow received by?

Two solid angles will be used for the remainder of the explanation:

The solid angle $d\Omega_1$ under which dS_2 is seen by O_1:

$$d\Omega_1 = \frac{dS_2 \cos\theta_2}{r^2} \qquad [1.22]$$

The solid angle $d\Omega_2$, under which dS_1 is seen by O_2:

$$d\Omega_2 = \frac{dS_1 \cos\theta_1}{r^2} \qquad [1.23]$$

The energy emitted by dS_1 and received by dS_2 is:

$$d\Phi_2 = L_T \cos\theta_1 \, d\Omega_1 \, dS_1 \qquad [1.24]$$

where L_T is the radiance of dS_1. This radiance is found elsewhere, defined by the preceding formula.

Replacing by its value, we find:

$$d\,\Phi_2 = L_T\,\cos\theta_1\,\frac{d\,S_2\cos\theta_2}{r^2}\,d\,S_1 \qquad [1.25]$$

which can be rewritten as:

$$d\,\Phi_2 = L_T\,d\,S_2\cos\theta_2\,\frac{d\,S_1\cos\theta_1}{r^2} = L_T\,d\,S_2\cos\theta_2\,d\,\Omega_2 \qquad [1.26]$$

We note that through this formula, if the surface was emitting the same luminance as before $d\,S_1$, L_{T1} the energy $d\,\Phi_2$ received by $d\,S_1$ would be equal to $d\,\Phi_1$. In fact:

$$d\,\Phi_1 = L_T\,d\,S_2\cos\theta_2\,d\,\Omega_2 = L_T\,d\,S_2\cos\theta_2\,\frac{d\,S_1\cos\theta_1}{r^2} = d\,\Phi_2 \qquad [1.27]$$

which forms Bouguer–Lambert's law.

The quantity $\dfrac{d\,S_1\,d\,S_2\cos\theta_1\cos\theta_2}{r^2}$ is often called the **optical range** determined by the surfaces $d\,S_1$ and $d\,S_2$.

NOTE.– This result is used particularly in establishing the Kirchhoff Law. In fact, two surfaces of the same emissivity, facing each other at the same temperature in an enclosure, are in thermodynamic equilibrium; they both send each other the same flow.

1.3.4. *Intensity*

When we have a body with small dimensions, we consider the flow $d\,\Phi$ that it emits over the whole surface, in a given direction. This energy flow $d\,\Phi$ is then defined in a small solid angle $d\,\Omega$ centered on the chosen direction. We then have to define monochromatic intensity $I(\theta,\lambda)$, by:

$$d^2\Phi = I(\theta,\lambda)\,d\,\Omega\,d\,\lambda = I_\lambda(\theta,)\,d\,\Omega\,d\,\lambda \qquad [1.28]$$

And a total intensity $I(\theta)$, by:

$$d\Phi = I(\theta)\, d\Omega \quad [1.29]$$

The curve $I = I(\theta)$ is often represented in a polar diagram. The curve obtained is then called an intensity indicator. In practice, applied to a light source (an ampoule, for example), it will give, in particular, the energy sent by this ampoule in all directions.

1.3.5. *Lambert's law – a surface's emissivity*

1.3.5.1. *Lambert's law*

Many light sources obey this law, which will prove very practical in applications. Moreover, all emitting bodies considered in our examples satisfy it.

A light **source** meets Lambert's law when its radiance does not depend on the angle of incidence, that is, when it is identical in all directions.

We therefore have:

$$\frac{d\, L_{T,\lambda}(\theta)}{d\theta} = 0 \quad [1.30]$$

Note that it can be written as:

$$d^3\,\Phi = L_{T,\lambda}\; d\,S \cos\theta\, d\lambda\, d\Omega \quad [1.31]$$

The dependence of $L_{T,\lambda}$ on θ disappears, but following a remark already made above, the dependence on $\cos\theta$ remains.

1.3.5.2. *A surface's emissivity*

In practice, it is important to know the energy sent by a surface in all directions. We should therefore integrate radiance over the whole half-space, which is situated above the surface considered.

This notion is particularly important. We will make intensive use of it in the examples used in these chapters on radiation.

In practice, we will often have to establish balances on a surface. To do this, it is practical to express the energy sent by a surface in all directions. To do this, we will integrate the expression of the flow emitted over a half-space, which is a solid angle

$$\Omega = \int_{half\ space} d\Omega = \pi \qquad [1.32]$$

This leads us to the definition, important for the remainder of the book, of a monochromatic emittance, and a total emittance of a surface. This emittance is defined locally on an elementary surface dS.

We will thus have to define **monochromatic emittance** $M_{T,\lambda}$, the expression of the flow emitted in the whole half-space:

$$d^2\Phi_\lambda = \int_{half\ space} d^3\Phi = d\lambda \int_{half\ space} L_{T,\lambda}(\theta)\, dS \cos\theta\, d\Omega = M_{T,\lambda}\, dS\, d\lambda \qquad [1.33]$$

Thus,

$$M_{T,\lambda} = \int_{half\ space} L_{T,\lambda}(\theta) \cos\theta\, d\Omega \qquad [1.34]$$

In the same way, we will have to define **the total emittance** M_T, the expression of the flow emitted by dS in the whole half-space:

$$d\Phi_\lambda = \int_{half\ space,\,\lambda} d^3,,\Phi = \int_{half\ space,\,\lambda} L_{T,\lambda}(\theta)\, dS \cos\theta\, d\lambda\, d\Omega = M_{T,\lambda}\, dS \qquad [1.35]$$

So,

$$M_T = \int_{half\ space} L_{T,\lambda}(\theta) \cos\theta\, d\Omega\, d\lambda \qquad [1.36]$$

We note that the integral in the whole half-space integrates the information on the dependence of the light with θ and $\cos\theta$.

We will not reproduce the resulting calculation here, which is purely geometrical by nature.

The result is particularly important, and we will use it implicitly or explicitly in the examples of a frequency application.

In fact, we find, for monochromatic emittance:

$$M_{T,\lambda} = \int_{half\ space} L_{T,\lambda}(\theta) \cos\theta \, d\Omega = \pi L_{T,\lambda}(\theta) \qquad [1.37]$$

and, integrated over all the wavelengths, we will find for the total emittance:

$$M_T = \int_{half\ space} L_{T,\lambda}(\theta) \cos\theta \, d\Omega d\lambda = \pi \, L_{T,\lambda}(\theta) \qquad [1.38]$$

This relationship, which results directly from the orthopty of the source, is also often called Lambert's law.

1.3.5.3. *Intensity indicator of a Lambertian source*

Another consequence of Lambert's law has implications for intensity. If we represent the intensity variation $I(\theta)$ with the incidence θ by vectors issuing from a single point, we trace what is called the intensity indicator. This type of diagram proves particularly useful in all problems involving light. In fact, it technically defines the source used.

For a Lambertian source, the intensity indicator is a circle.

NOTE.–

Lambert's name is linked to three different laws that should not be confused:

– Bouguer–Lambert's law on optical range;

– Lambert's law expressing the orthopty of a source and above all, linking emittance and radiance;

– Beer–Lambert's law, relating to the absorption by transparent or translucid bodies, which we will briefly present in Chapter 3.

2

Calculating Luminances

2.1. Introduction

In terms of radiation, emission from a material, body or medium is calculated with reference to that of an ideal body, the "**black body**". We will define and introduce this very particular model, which led to a number of publications and lay at the root of a "new physics". More modestly here, we will take it as a vital basis for determining luminances.

The luminance of the black body will be given by **Planck's law (section 2.2)**. It will enable us to determine the total emittance of a black body using the **Stefan–Boltzmann law (section 2.3)**; **Wien's law (section 2.4)** we will give its maximum emission. In practice, it will be very useful to be able to easily calculate the emission of such a black body in a finite spectral interval. In $G_{0-\lambda T}$ functions, there will be a vital tool for doing this **(section 2.5)** We will therefore be able, using the definition of emissivity, to calculate the energy luminance, then the emittance, which is in fact often more practical for any material **(section 2.6)**.

We will note in passing that, following many other works on this matter, this book adopts an approach practically opposed to the traditional approach. The experimental approach in fact preceded Planck's theoretical modeling by several decades. Nonetheless, this will be the point of departure below to establish our radiation calculating tools.

2.2. The black body: concept, luminance, Planck's law and approximations

2.2.1. *Paradoxically, the black body is defined with reference to its absorption*

A black body is an "idealized device" which absorbs all the light it receives. This concept was presented by Kirchhoff in 1862.

Kirchhoff's law will show that this **black body** is consequently the body that emits the most radiation at a given temperature. Qualitatively, this can be stated: a body that always absorbed without re-emitting would heat up continuously until evaporation or sublimation. It should therefore re-emit continuously to remain in thermodynamic equilibrium. And whatever absorbs more re-emits more, etc.

Below **(section 2.5)**, we will see the Kirchhoff laws that directly link emission and absorption in more detail.

We suggest that the curious reader goes far beyond these summaries. The reader will then see that, as far as this chapter is concerned, the notion of thermodynamic equilibrium has essentially guided the grandfathers of the discipline.

Different devices that meet the definition of a "black body"

A closed cavity in which a small hole is drilled is an excellent physical model of a black body and forms an excellent approximation of it: the light comes in, undergoes internal reflections and statistically it will not escape. An oven with a small hole also serves this purpose. Kirchhoff established using thermodynamics reasoning that the emission of a black body is known and calculable when its temperature is known, which becomes the sole parameter of this emission. The "black body" oven also becomes, in a laboratory, a "calibrated luminance source for photometric measures in various domains". This experimental model also served Wien (displacement law) to establish experimentally a law that we could also deduce from Planck's law.

To create "an idea", consider **the front of a house with an open window**: the light rushes into the room and will not leave, hence, the appearance of a "black spot" on a photograph of the house (of course, there must not be a mirror on the wall opposite the window, and within the range of the camera!).

Other physical emitters can be assimilated to black bodies:

A plate covered in soot due to smoke (a carbon deposit resulting from smoke from a "rich" flame) is a very valuable approximation.

Astrophysics provides the example of the **"continuous" spectrum of** moderate temperature **stars**. Our "Sun", at least its surface, behaves with fairly good approximation like a black body with an estimated temperature between 5,700 and 5,900°C. This approximation results from the fact that the spectrum is not perfectly monotonous in its left part.

NOTE.– Observed from the Earth, **the solar spectrum** also shows "absorption rays" (**Fraunhofer rays**). This temperature results from the surface of the Sun and has no common measure with the star's internal temperature that is the location of nuclear fission reactions. We will return with these two remarks to the field of radiation transfer in transparent and absorbent gaseous atmospheres, which will not be studied in this book.

The **clusters of interstellar dust** also form a black body at hundreds of Kelvins.

The **cosmological diffuse background**, which is supposed to result and date from the "Big Bang", behaves practically perfectly as a black body at 2,728 K.

2.2.2. *Black body luminance*

To calculate the luminance of a body at a given temperature, we should therefore know the luminance of a particular type of material called the black body. We will see how in **section 2.6**.

This luminance was first known experimentally, through the works of Stefan, Wien and Tyndall, to cite the most well known. The difficulties posed by modeling these phenomena had considerable influence on the evolution of physics. The theory suggested by Rayleigh and Jeans resulted in failure: a continuous spectrum is obtained that gives a good approximation of the experiment with large wavelengths (on the lower right-hand side, we will see the black body spectrum) by a luminance increasing indefinitely when the wavelength decreases towards $\lambda = 0$. The model then remains valid in the infrared, but deviates completely in ultraviolet and the highest frequencies (i.e. the lowest wavelengths, X rays, γ rays, etc.). This will be the source for the physics of an "ultraviolet catastrophe"; a catastrophe whose physics was discovered by Planck in a publication dated December 14, 1900 on a model based on the hypothesis that the spectrum was discontinuous, where the light energy was distributed by a very large number of grains of value $h\nu$, where ν is the frequency of a radiation.

Initially designed as a mathematical artifice, this concept went on to have a great future under the name "photon", given by a certain Albert Einstein.

As we can see, our chapter is a good summary of the great names in physics. Planck won the Nobel Prize for this work in 1918 and Einstein in 1905. The list of awards may be surprising.

Experiment and theory therefore now agree in giving a **spectral distribution of the luminance of a black body**:

$$L^0(\lambda,T) = L^0_{\lambda,T} = \frac{c_1}{\lambda^5}\frac{1}{\exp\left(\frac{C_2}{\lambda T}\right) - 1} \quad [2.1]$$

With two constants:

$$c_1 = 2hc^2 = 1.19.10^{-16}\, W.m^2 \quad [2.2]$$

$$C_2 = \frac{hc}{k} = 1.4388.10^{-2}\, m.K = 14388\, \mu m.K \quad [2.3]$$

We note the decision to use uppercase and lowercase in writing these constants. c, h and k are respectively:

– the celerity of light in a vacuum:

$$c = 299792458\, m.s^{-1} \approx 3.10^8\, m.s^{-1} \quad [2.4a]$$

– Planck's constant h, which defines the energy $h\nu$ of the photon:

$$h = 6.62617.10^{-34}\, J.s \quad [2.4b]$$

– Boltzmann's constant:

$$k = 1.38066.10^{-23}\, J.K^{-1} \qquad [2.4c]$$

NOTE.– Boltzmann's constant is fundamental in statistical gas theory. It makes it possible to reconstruct, using the Avogadro number N, the constant R of the ideal gasses:

$$R = k\, N \qquad [2.5]$$

The function $L^0_{\lambda,T}$ extends over a large number of orders of grandeur. It should be represented on a log–log plot. It is difficult to represent in its full extent. We note that it has null values at the value limits $\lambda = 0$ and $\lambda \to \infty$.

Figure 2.1 gives a general impression of this function for some temperatures. We note that in many practical cases, which we will use, the Stefan–Boltzmann law will make it possible to circumvent direct calculation of these values.

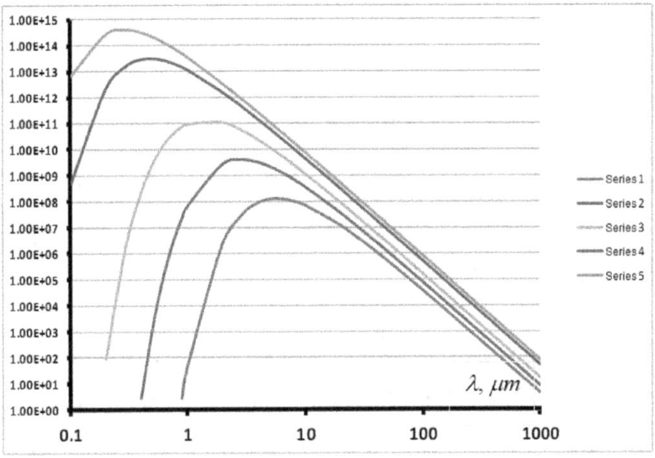

Figure 2.1. *General appearance of the luminance $L^0_{\lambda,T}$ of the black body, at some temperatures. Series 1: $10000\,K$; series 2: $6000\,K$; series 3: $2000\,K$; series 4: $1000\,K$; series 5: $500\,K$. For a color version of this figure, see www.iste.co.uk/ledoux/heat2.zip*

It will also be noted that at each temperature, the luminance of the black body has a maximum that lies at a wavelength that varies with this temperature. Wien's (first) law will enable us to determine this maximum.

2.2.3. *Emittance from the black body*

As in many publications, everything relating to the black body is written with an upper "0". The luminance is thus written $L^0_{\lambda,T}$.

NOTE.– We have chosen to let the two notations $L^0\,(\lambda,T)$ and $L^0_{\lambda,T}$ co-exist. The first recalls the mathematical dependency [relationship] and the second is more a notation for practical use. The curious reader or anyone wishing to take this subject further will see that scientific literature, manuals or articles accept some diversity in notation.

In the examples, we will work for the most part in terms of emittance that expresses, remember, the energy emitted in a small interval around a wavelength in all directions.

The black body obeys Lambert's law. This law is often applied in terms of monochromatic emittance; we will therefore write here:

$$M^0\,(\lambda,T) = M^0_{\lambda,T} = \pi\, L^0\,(\lambda,T) \qquad [2.6]$$

There again, we use the upper index "0". We therefore have:

$$M^0\,(\lambda,T) = M^0_{\lambda,T} = \frac{C_1}{\lambda^5}\frac{1}{\exp\left(\dfrac{C_2}{\lambda T}\right) - 1} \qquad [2.7]$$

We note that the constant C_2 has not varied, but that multiplication by π has changed the constant c_1, which changes the status and becomes $C_1 = \pi\, c_1$ upper case as C_2:

$$C_1 = 2\pi h c^2 = 3.74.10^{-16}\, W.m^2 \qquad [2.8]$$

$$C_2 = \frac{hc}{k} = 1.4388.10^{-2}\, m.K = 14388\,\mu m.K \qquad [2.9]$$

2.2.4. *Approximations of the luminance of the black body*

For particular values of the wavelength, this law can be simplified with the help of two useful approximations. **These approximations are addressed to ranges of "small" or "large" wavelengths**, that is, ultraviolet and infrared. "Small" or "large" has no meaning without a reference.

a) Rayleigh–Jeans approximation

Corresponding to the modeling mentioned above, the **Rayleigh–Jeans law is valid at long wavelengths**, that is, at small frequencies. This law in fact refers to Jeans' first model, based on continuous electromagnetic oscillators, which led to a divergent process (infinite emission at high frequencies).

Consider:

$$\lambda >> \frac{hc}{kT} \quad [2.10]$$

which is equivalent to:

$$\lambda T >> \frac{hc}{k} \quad [2.11]$$

Planck's formula is simplified to:

$$L^0(\lambda,T) \approx \frac{C_1}{C_2}\lambda^{-4}T = \frac{2kcT}{\lambda^4} \quad [2.12]$$

The relative error produced by this approximation can be evaluated by:

$$\frac{\Delta L^0}{L^0} \approx \frac{-C_2}{2\lambda T} \quad [2.13]$$

This law is clearly not valid at small wavelengths, as it diverges for $\lambda = 0$.

b) Wien's approximation

This approximation is valid at **small wavelengths** (high frequencies).

$$\text{If } \lambda T \approx \varepsilon \qquad [2.14]$$

then Planck's formula is simplified to:

$$L^0(\lambda,T) \approx c_1\,\lambda^{-5}\exp\left(\frac{-C_2}{\lambda T}\right) \qquad [2.15]$$

The relative error produced by this approximation can be evaluated by:

$$\frac{\Delta L^0}{L^0} \approx \exp\left(\frac{-C_2}{\lambda T}\right) \qquad [2.16]$$

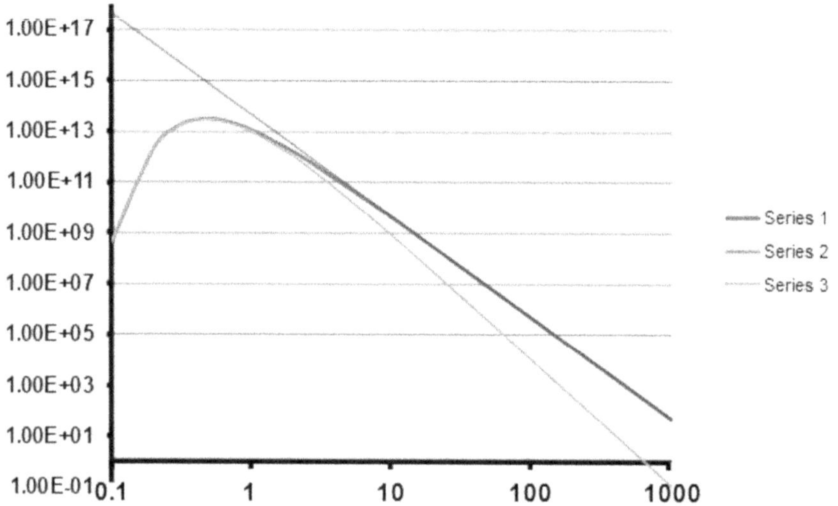

Figure 2.2. *Comparison of the exact luminance $L^0_{\lambda,T}$ of the body with both approximations. Blue line, series 1: exact luminance $L^0_{\lambda,T}$. Red line, series 2: Rayleigh–Jeans approximation. Illustration of the "energy catastrophe". Green line, series 3: Wien approximation. For a color version of this figure, see www.iste.co.uk/ledoux/heat2.zip*

This approximation is particularly effective for small wavelengths; in this case, the exponential $\exp\left(\frac{C_2}{\lambda T}\right)$ is large before 1 and $\frac{1}{\exp\left(\frac{C_2}{\lambda T}\right) - 1}$ is very well approximated by

$$\frac{1}{\exp\left(\frac{C_2}{\lambda T}\right)} = \exp\left(\frac{-C_2}{\lambda T}\right) \qquad [2.17]$$

Both of these approximations are compared to the exact luminance in Figure 2.2.

2.2.5. *Writing the luminance in terms of frequency*

This note is made on the assumption that the reader will consult the literature. The calculation made here in a specific case can be added to the general considerations in **section 1.3.2**.

The flow expression $d^3\Phi$ can also be written in terms of the spectral interval indicated in frequencies ν. A form of writing luminance $L_{T,\nu}$ is therefore introduced.

We will connect

$$d^3\,\Phi = L_{T,\nu}\,(\theta)\,d\,S\cos\theta\,d\nu\,d\,\Omega \qquad [2.18]$$

The expressions will be linked by writing the integrated radiated energy across the whole spectrum.

The expression $L_{T,\nu}\,(\theta)$ will then be different from that of $L_{T,\lambda}\,(\theta)$.

We thus determine the frequency expression of the black body luminance

$$L^0_{\lambda,T} = \frac{c_1}{\lambda^5}\frac{1}{\exp\left(\frac{C_2}{\lambda T}\right) - 1} = \frac{2hc^2}{\lambda^5}\frac{1}{\exp\left(\frac{C_2}{\lambda T}\right) - 1} \qquad [2.19]$$

Integrated over the whole spectrum, the two forms of the luminance should therefore meet:

$$\int_0^\infty L^0_{\lambda,T}\,d\lambda=\int_0^\infty \frac{c_1}{\lambda^5}\frac{1}{\exp\left(\dfrac{C_2}{\lambda T}\right)-1}d\lambda=\int_0^\infty \frac{2hc^2}{\lambda^5}\frac{1}{\exp\left(\dfrac{C_2}{\lambda T}\right)-1}d\lambda=\int_0^\infty L^0_{\nu,T}\,d\nu \quad [2.20]$$

Noting that:

$$\lambda=\frac{c}{\nu} \quad [2.21]$$

and therefore:

$$d\lambda=-\frac{c}{\nu^2}d\nu \quad [2.22]$$

the integral can be rewritten in λ in the following form, remembering that $\lambda=\dfrac{c}{\nu}$:

$$\int_0^\infty \frac{2hc^2}{\lambda^5}\frac{1}{\exp\left(\dfrac{C_2}{\lambda T}\right)-1}d\lambda=\int_0^\infty 2hc^2\left(\frac{\nu}{c}\right)^5\frac{1}{\exp\left(\dfrac{hc}{k}\dfrac{\nu}{cT}\right)-1}\frac{-c}{\nu^2}d\nu \quad [2.23]$$

$$\int_0^\infty \frac{2hc^2}{\lambda^5}\frac{1}{\exp\left(\dfrac{C_2}{\lambda T}\right)-1}d\lambda=-\int_0^\infty 2\frac{h\nu^3}{c^2}\frac{1}{\exp\left(\dfrac{h\nu}{kT}\right)-1}d\nu \quad [2.24]$$

and:

$$-\int_0^\infty 2\frac{h\nu^3}{c^2}\frac{1}{\exp\left(\dfrac{h\nu}{kT}\right)-1}d\nu=\int_0^\infty L_{\nu,T}\,d\nu \quad [2.25]$$

NOTE.– The sign "–" represents an artifact: in fact, integration on increasing values of "λ" corresponds to an integration on decreasing values of "ν".

Mathematically, the integrals obtained are of different signs. This type of reasoning is found in thermodynamics when calculating the integral of terms like "*pdV*". The sign of the calculated work may be significant. Here, in terms of equality of two calculations of energy quantity, we can discard it.

And by identification we find:

$$L^0_{\nu,T} = \frac{2h\nu^3}{c^2}\frac{1}{\exp\left(\dfrac{h\nu}{kT}\right) - 1} \qquad [2.26]$$

which is the starting point for many scientific publications. It is therefore important to know this expression when exploring the literature.

NOTE.– Readers familiar with statistical thermodynamics will have recognized in the term $\frac{h\nu}{kT}$ the energy ratio of the photon $\frac{h\nu}{kT}$ to the Boltzmann term $\frac{h\nu}{kT}$. This is not surprising, since Planck's rationale was based exactly on a statistical calculation (Boltzmann was a pupil of Planck who contributed to this theory).

2.3. Stefan–Boltzmann law

2.3.1. *Establishing the law*

In the subject of radiation, the main laws have been expressed on an experimental basis, before being integrated in a synthetic theoretical vision based on Planck's theory.

The Stefan–Boltzmann law was posed experimentally through the work of Joseph Stefan in 1879 using Tyndall's experimental data. The theoretical foundations were posed in the context of thermodynamics by one of Stefan's doctoral students, Ludwig Boltzmann (1844–1906), in 1884.

We will establish it here as a result of the luminance data, or more precisely the emittance from a black body surface.

The total emittance of a black body in all directions and across the whole (infinite) wavelength spectrum is calculated by integrating the total monochromatic emittance of $\lambda = 0$ to $\lambda \to \infty$

$$M^0(T) = \pi L^0(\lambda, T) = \int_0^\infty M^0(\lambda, T)\, d\lambda \qquad [2.27]$$

$$M^0(T) = \int_0^\infty \frac{C_1}{\lambda^5} \frac{1}{\exp\left(\frac{C_2}{\lambda T}\right) - 1} d\lambda \qquad [2.28]$$

This integral is transformed easily by carrying out a change of variable:

$$u = \lambda T \qquad [2.29]$$

$$\text{So, } du = T d\lambda \qquad [2.30]$$

and:

$$M^0(T) = \int_0^\infty C_1 \frac{T^5}{u^5} \frac{1}{\exp\left(\frac{C_2}{u}\right) - 1} \frac{du}{T} \qquad [2.31]$$

Posing

$$\xi = \frac{C_2}{u} \qquad [2.32]$$

We obtain

$$M^0(T) = \left[C_1 \int_0^\infty \frac{\xi^5}{C_2^5} \frac{1}{\exp(\xi) - 1} \frac{C_2\, d\xi}{\xi^2} \right] T^4 = \sigma T^4 \qquad [2.33]$$

σ is a constant equal to:

$$\sigma = \frac{C_1}{C_2^4} \int_0^\infty \frac{\xi^3}{\exp(\xi) - 1} d\xi \qquad [2.34]$$

The constant σ contains an integral that cannot be calculated simply. Its value is:

$$\int_0^{\infty} \frac{\xi^3}{\exp(\xi)-1} d\xi = \frac{\pi^4}{15} \qquad [2.35]$$

Considering the values of

$$C_1 = 2\pi hc^2 \text{ and } C_2 = \frac{hc}{k}, \qquad [2.36]$$

σ results in

$$\sigma = \frac{2\pi^5 k^4}{15c^2 h^3} = 5,67.10^{-8}\ W.m^{-2}K^{-1} \qquad [2.37]$$

And the total emittance M_T^0, that is, the energy emitted per unit of time and surface by a plane black body at the absolute temperature is given by the law known as the Stefan–Boltzmann law:

$$M_T^0 = \sigma T^4 \qquad [2.38]$$

$$\sigma = 5,67.10^{-8}\ W.m^{-2}K^{-1} \qquad [2.39]$$

2.3.2. *A direct application*

We will make extensive use of this law here. There will be an example of it here and many others will be found in Chapter 5.

EXAMPLE 2.1.– Solar illumination

Suppose that the Sun radiates like a black body corps at temperature T_S = 5900 K.

We wish to evaluate the solar illumination over $1\,m^2$ of the Earth's surface.

1) Assimilate the radiative exchange between the Sun and one m² of the Earth at an exchange between two small surfaces. Justify this approximation.

2) Evaluate the radiated flow by the Sun over one m^2 of Earth. Does this value seem compatible to you based on what you have already learned about terrestrial insolation?

***Data*:** A ray of the Sun: $R_S = 700\,000\,km$; distance Earth/Sun: $D_{TS} = 150\,000\,000\,km$

SOLUTION.–

1) The ratio of the ray of the Sun to the distance between the Sun and the Earth is $\alpha = \dfrac{150\,000\,000}{700\,000} = 214.3$. We could paradoxically consider the surface of the Sun and one square meter of Earth considered as basic surfaces $d\,S_1$ and $d\,S_2$.

With

$$d\,S_1 = \pi\left(7.10^8\right)^2 = 1.54.10^{18}\,m^2 \text{ and } d\,S_2 = 1\,m^2 \qquad [2.40]$$

We will consider one m² of Earth placed at the equator, so normal to the axis of the Sun and the Earth.

2) The flow exchanged will therefore be given by the Bouguer–Lambert formula:

$$\Phi = L_T^0\,\frac{d\,S_1\,d\,S_2}{D_{TS}^2} = \frac{\sigma T^4\,d\,S_1\,d\,S_2}{\pi\,D_{TS}^2} \qquad [2.41]$$

where L_T^0 is the solar radiation, linked to its emittance by:

$$L_T^0 = \frac{M_T^0}{\pi} = \frac{\sigma T^4}{\pi} = \frac{5.67.10^{-8} * 5900^4}{\pi} = 2.19.10^7\,W.m^{-2} \qquad [2.42]$$

$$\Phi = L_T^0 \frac{dS_1\,dS_2}{D_{TS}^2} = \frac{2.19.10^7 * 1.54.10^{18} * 1}{\left(150.10^6\right)^2} \qquad [2.43]$$

$$\Phi = 1498\ W.m^{-2} \qquad [2.44]$$

NOTE.– This illuminance per surface unit refers to a surface normal to the axis of the Sun and the Earth. An average illuminance can be calculated by noting that the total flow Φ_T arriving on Earth will cover a surface $S_C = \pi R_T^2$, where R_T is the radius of the Earth.

$$\text{So } \Phi_T = 1498 * \pi R_T^2 \qquad [2.45]$$

However, the total area on which this illuminance shines is a hemi-sphere, which is

$$S_E = 2\pi R_T^2 \qquad [2.46]$$

The average flow per surface unit illuminating the Earth during daylight will therefore be:

$$\Phi_T = \frac{1498 * \pi R_T^2}{2\pi R_T^2} = 749\ W.m^{-2} \qquad [2.47]$$

We note that some climate models give an average of the illuminance over 24 hours, decreasing this value by two considering the illuminance over the whole Earth's surface, equal to $4\pi R_T^2$, which is a flow

$$\Phi_T = 156\ W.m^{-2} \qquad [2.48]$$

We finally note that this illuminance, considering the incidence of the solar rays ranging from $0°$ to $90°$, will be heterogeneous.

2.4. Wien's laws

The luminance (like the emittance) of the black body is a curve that has a maximum. This maximum corresponds to the zone where the maximum energy is emitted. Considering the number of orders of size over which Planck's law extends, this maximum is physically significant. Both of Wien's laws give its characteristics. Wien's first law, or displacement law, by far the most common, indicates where the maximum is found; the second gives its intensity.

2.4.1. *Wien's displacement law*

For a given temperature, Planck's law should have a wavelength for which the luminance of the black body is maximum. This wavelength λ_m depends on the temperature. This is an important parameter as it determines in particular the order of size of the wavelengths where the emissivity is maximum. Under some conditions, it could determine the color of a solid body at a given temperature.

The monochromatic luminance of a black body is given by Planck's law:

$$L^0_{\lambda,T} = L^0_\lambda(T) = L^0(\lambda,T) = \frac{C_1}{\lambda^5}\frac{1}{\exp\left(\frac{C_2}{\lambda T}\right)-1} \quad [2.49]$$

with:

$$C_1 = 2hc^2 = 1.19.10^{-16}\,W.m^2 \quad [2.50]$$

$$C_2 = \frac{hc}{k} = 1.4388.10^{-2}\,m.K = 14388\,\mu m.K \quad [2.51]$$

NOTE.– We have again decided to reveal the three alternative notations for luminance, $L^0_{\lambda,T}$, $L^0_\lambda(T)$ or $L^0(\lambda,T)$ which can appear in different works. Each notation can be adapted to a different perspective. Here, for a given temperature T, we seek the wavelength λ giving the maximum of L^0. The notation $L_\lambda^0(T)$ is therefore best adapted. To seek this maximum, we will express the luminance as a function of the parameter contained in the exponential, through a change of variable:

$$u = \frac{C_2}{\lambda T} \qquad [2.52]$$

$$L^0(\lambda,T) = L^0(\lambda,T) = \frac{C_1}{\lambda^5}\frac{1}{\exp(u)-1} = \frac{C_1 T^5}{C_2^5}\frac{u^5}{\exp(u)-1} \qquad [2.53]$$

For a given temperature T, the curve $L^0(u,T) = f(u)$ is an increasing decreasing function. Therefore, the maximum luminance will be observed for a value of $u = \lambda T$ such that:

$$\frac{\partial}{\partial u}\frac{u^5}{\exp(u)-1} = 0 \qquad [2.54]$$

Developing the calculation:

$$\frac{\partial}{\partial u}\frac{u^5}{\exp(u)-1} = \frac{5u^4(\exp(u)-1) - u^5\exp(u)}{\left[\exp(u)-1\right]^2} \qquad [2.55]$$

The denominator is positive everywhere for $u > 0$, and the maximum sought will correspond to:

$$5u^4(\exp(u)-1) - u^5\exp(u) = u^4\left[5(\exp(u)-1) - u\exp(u)\right] = 0 \qquad [2.56]$$

which is:

$$(5-u)\exp(u) - 5 = 0 \qquad [2.57]$$

or indeed:

$$\exp(u) = \frac{5}{5-u} \qquad [2.58]$$

This equation can easily be solved digitally; the result obtained is:

$$u = 0.96511 \quad [2.59]$$

This leads to a demonstration that the luminance maximum at the given temperature T is obtained for a wavelength λ_{max} such that:

$$\lambda_{max} T = \frac{C_2}{u} = \frac{1.4388;10^{-2}}{4.96511} = 2.8978.10^{-3}\ m.K \quad [2.60]$$

which established the first Wien, also known as Wien's displacement law:

$$\lambda_{max} T = 2.8978.10^{-3}\ m.K \quad [2.61]$$

NOTE.– In practice, the wavelengths are often expressed in μm (microns). Wien's law is written correctly above in the SI system.

It can be rewritten in the form:

$$\lambda_{max} T = 2897.8\,\mu.K \quad [2.62]$$

where T is expressed in Kelvin and λ in micrometers.

It will be noted that this formula, in current use, will give the result for λ_{max} in micrometers.

2.4.2. *Wien's second law*

Wien's second law, expresses the value of the maximum determined by the displacement law. It is enough to carry the value of the maximum emission wavelength in the expression of the spectral emittance of the black body.

The value of the emittance at this wavelength λ_{max} is given by

$$M^0(\lambda_{max}) = BT^5 \quad [2.63]$$

$$B = 1.28.10^{-5}\ W.m^{-3}.K^{-5} \quad [2.64]$$

These laws are also in practice applicable to the emittance of the body because of the relationship:

$$M_{\lambda,T}^{0} = \pi L_{\lambda}^{0}(T) \tag{2.65}$$

$$M_{\lambda,T}^{0} = \pi L^{0}(\lambda,T) = \frac{C_1}{\lambda^5}\frac{1}{\exp\left(\frac{C_2}{\lambda T}\right)-1} \tag{2.66}$$

2.4.3. *Greenhouse effect*

Wien's law enabled us to better understand the phenomenon known in the media as the greenhouse effect, a term that is actually rather inaccurate (see below).

The system is simple in its principle. We consider a surface situated on the Earth's surface.

It receives solar radiation that is modeled by a black body in the region of $5900\,K$. The Earth, in the simple model currently adopted (which is enough for teaching purposes), is assimilated to a black body, we say at $300\,K$. Wien's law shows that most incident and emitted radiations are centered on radically different wavelengths:

For the Sun:

$$\lambda_{S\max} = \frac{2898}{5900} = 0.491\,\mu m\text{, which is in the visible spectrum} \tag{2.67}$$

For the Earth:

$$\lambda_{E\max} = \frac{2898}{300} = 9.66\,\mu m\text{, which is in infrared.} \tag{2.68}$$

Figure 2.3 shows the two corresponding spectral energy distributions.

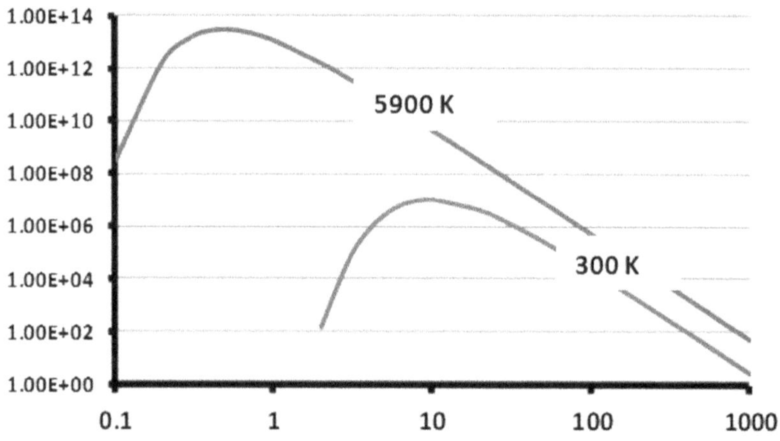

Figure 2.3. *Luminances compared for the Sun and the Earth, considered as a black body. For a color version of this figure, see www.iste.co.uk/ledoux/heat2.zip*

If, between the Sun and the Earth, we place a selective transmission material, which is opaque for infrared, the radiation re-emitted by the Earth is blocked and is responsible for heating the atmosphere above the ground.

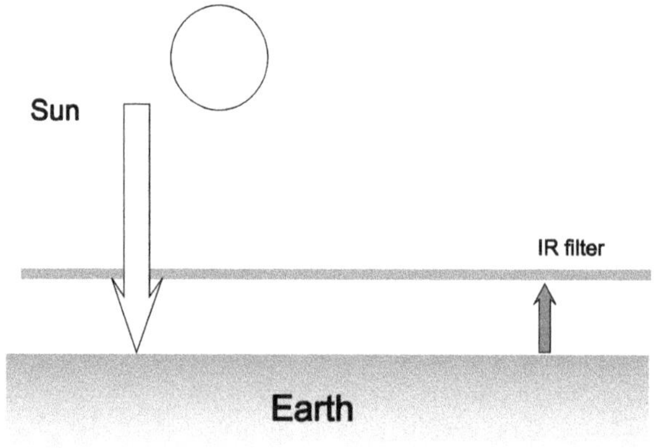

Figure 2.4. *Principle of radiative forcing. For a color version of this figure, see www.iste.co.uk/ledoux/heat2.zip*

On a lesser scale, plate glass is used for horticulture, and indeed for vegetables. The phenomenon is thus invoked to explain the greenhouse effect.

At atmospheric scale, the presence of absorbent gas in infrared retains the heat radiated by the Earth and the trend for it to increase the ambient temperature. These gasses, called "greenhouse gasses", are naturally present in the atmosphere: H_2O, CO_2, CH_4, the most present (water represents more than 70% of the effect), as well as N_2O (nitrous oxide) and O_3 (ozone). Their effect is therefore beneficial: it is often calculated by a "simple" model that the ambiance would be $T = -18$ °C without the "greenhouse effect".

But if, in the wake of human activity (industry, transport, domestic heating, agriculture, etc.), their concentration, especially the carbon dioxide of "anthropic" origin, is becoming excessive, then global warming is becoming immoderate too. It is beyond the scope of this book to analyze the intricacies of the scientific and political debates that follow.

The term "greenhouse effect", which we used for a time because of its use in the media, is incorrect. It is better to use the term "**radiative forcing**". In fact, the way a greenhouse works results essentially from thermal convection and not from the blocking of infrared radiation by glass. Robert Williams Wood demonstrated this in 1890, by replacing the glass in a greenhouse model with a sort of rock salt, a material that remains permeable to infrared (in IR optics, we use sodium chloride lenses).

NOTE.– Topical concerns on the effects of gasses of anthropic origin mask, in the public consciousness, another threat of thermal origin in a very distant future. At the geological scale, the ambient temperature undergoes strong oscillations. On cycles in the order of 100,000 years, the "mild" climate that we know will last 10,000 to 20,000 years. Our current "temperate" ***Holocene has lasted only 11,500 years***.

These oscillations should be explained by the remarks of Milutin Milankovic on the variations in orbit and incline on the Earth's rotation axis. Assuming that we do return to a "normal" rate of emissions, the future of the Earth has already been decided. Too often, we forget that 15,000 years ago, reindeer were hunted in the south of France!

And we want to forget that in a few thousand years, glaciation will return. It should be hoped that the order of thermal science will have held firm, and that more generally, human technology will have properly integrated this threat. Otherwise, what will be the destiny of an already overpopulated humanity that will have long ago exhausted its fossil and probably its fissile fuels?

This problem is evidently beyond the scope of this book, but it seems important, from a "historical" and "cultural" perspective, to remark on it.

2.5. Fraction of the total emittance of a black body radiated in a spectral band

2.5.1. *An important tool:* $G_{0-\lambda T}$ *functions*

In practice, it is necessary to determine the energy emitted (or absorbed) by a solid body in a wavelength band that does not cover the whole spectrum. Many bodies do not emit across the whole spectrum. This will also be the case when using filters. Moreover, it will be demonstrable that most of the emission lies on a band close to the maximum of Planck's law.

The following establishes a tool that will be equally useful in processing many problems, especially those involving a gray body.

Here, we will address the problem in terms of emittance.

The emittance of a black body integrated between two wavelengths λ_1 and λ_2 can be written as:

$$M^0_{\lambda_1-\lambda_2,T} = \int_{\lambda_1}^{\lambda_2} M^0_{\lambda,T}\, d\lambda \qquad [2.69]$$

This integral will be broken down into two parts:

$$\int_{\lambda_1}^{\lambda_2} M^0_{\lambda,T}\, d\lambda = \int_0^{\lambda_2} M^0_{\lambda,T}\, d\lambda - \int_0^{\lambda_1} M^0_{\lambda,T}\, d\lambda \qquad [2.70]$$

The Stefan–Boltzmann law, established above, enables us to write, across the whole spectrum, that is, for $\lambda_1 = 0$ and $\lambda_2 \to \infty$

$$M^0_{0-\infty,T} = \int_0^\infty M^0_{\lambda,T}\, d\lambda = M^0 = \sigma T^4 \qquad [2.71]$$

where

$$\sigma = 5.67.10^{-8}\, W.m^{-2}.K^{-4} \qquad [2.72]$$

The fraction of energy radiated by a black body between λ_1 and λ_2 will therefore be given by a fraction $F_{\lambda_1-\lambda_2,T}$:

$$F_{\lambda_1-\lambda_2,T} = \frac{\int_{\lambda_1}^{\lambda_2} M^0_{\lambda,T}\, d\lambda}{\int_0^\infty M^0_{\lambda,T}\, d\lambda} \qquad [2.73]$$

$$= \frac{1}{\sigma T^4} \int_{\lambda_1}^{\lambda_2} \frac{C_1\, d\lambda}{\lambda^5 \left[\exp\left(\frac{C_2}{\lambda T} \right) - 1 \right]} \qquad [2.74]$$

with

$$C_1 = 2\pi h c^2 = 3.74.10^{-16}\, W.m^2 \qquad [2.75]$$

$$C_2 = \frac{hc}{k} = 1.4388.10^{-2}\, m.K = 14388\, \mu m.K \qquad [2.76]$$

Note that this function depends on λ_1, λ_2 as well as on T.

The fraction of energy radiated by a black body between 0 and λ will therefore be given by a fraction $F_{0-\lambda}$:

$$F_{0-\lambda,T} = \frac{\int_0^\lambda M^0_{\lambda,T}\, d\lambda}{\int_0^\infty M^0_{\lambda,T}\, d\lambda} \qquad [2.77]$$

$$= \frac{1}{\sigma T^4} \int_0^{\lambda} \frac{C_1 \, d\lambda}{\lambda^5 \left[\exp\left(\frac{C_2}{\lambda T} \right) - 1 \right]} \quad [2.78]$$

with

$$C_1 = 2\pi hc^2 \quad \text{and} \quad C_2 = \frac{hc}{k} \quad [2.79]$$

We evidently have:

$$F_{\lambda_1 - \lambda_2, T} = \frac{1}{\sigma T^4} \int_{\lambda_1}^{\lambda_2} \frac{C_1 \, d\lambda}{\lambda^5 \left[\exp\left(\frac{C_2}{\lambda T} \right) - 1 \right]}$$

$$= \frac{1}{\sigma T^4} \int_0^{\lambda_2} \frac{C_1 \, d\lambda}{\lambda^5 \left[\exp\left(\frac{C_2}{\lambda T} \right) - 1 \right]} - \frac{1}{\sigma T^4} \int_0^{\lambda_2} \frac{C_1 \, d\lambda}{\lambda^5 \left[\exp\left(\frac{C_2}{\lambda T} \right) - 1 \right]} \quad [2.80]$$

which is rewritten as:

$$F_{\lambda_1 - \lambda_2, T} = F_{0-\lambda_2 T} - F_{0-\lambda_1 T} \quad [2.81]$$

The functions $F_{\lambda_1 - \lambda_2, T}$ can therefore be calculated from a set of functions $F_{0-\lambda, T}$. Since it is not immediate, the calculation of $F_{0-\lambda, T}$ integrals should in practice involve a digital method. In practice, we should therefore tabulate them. At this level, there is a difficulty, since the table of {? } should have two inputs, λ and T.

A mathematical stratagem will enable us to remove this difficulty and to reduce the tabulation to a single input. In fact, it is possible to show that $F_{0-\lambda, T}$ depends only on λT.

By introducing

$$u = \lambda T \quad [2.82]$$

$$F_{0-\lambda} = \frac{1}{\sigma T^4} \int_0^{\lambda} \frac{C_1 d\dfrac{u}{T}}{\left(\dfrac{u}{T}\right)^5 \left[\exp\left(\dfrac{C_2}{u}\right) - 1\right]} = \frac{C_1}{\sigma} \int_{u=0}^{u=\lambda T} \frac{du}{u^5 \left[\exp\left(\dfrac{C_2}{u}\right) - 1\right]} \quad [2.83]$$

with:

$$C_1 = 2\pi hc^2 \quad \text{and} \quad C_2 = \frac{hc}{k} \quad [2.84]$$

$F_{0-\lambda,T}$ will thus be tabulated as a function of the variable $u = \lambda T$. It will then be distinguished by renaming it

$$G_{0-\lambda,T} = F_{0-\lambda,T} : \quad [2.85]$$

$$F_{0-\lambda,T} = G_{0-\lambda T} = \frac{C_1}{\sigma} \int_{u=0}^{u=\lambda T} \frac{du}{u^5 \left[\exp\left(\dfrac{C_2}{u}\right) - 1\right]} \quad [2.86]$$

$$F_{0-\lambda,T} = G_{0-\lambda T} = 6.4905.10^{-9} \int_{u=0}^{u=\lambda T} \frac{du}{u^5 \left[\exp\left(\dfrac{1,4388.10^{-2}}{u}\right) - 1\right]} \quad [2.87]$$

To calculate the fraction of the **total emittance radiated between the wavelengths λ_1 and λ_2, $F_{\lambda 1-\lambda 2}$ will be calculated by the difference of the two functions $G_{0-\lambda T}$**

$$F_{\lambda_1-\lambda_2,T} = G_{\lambda_1 T-\lambda_2 T} = \frac{C_1}{\sigma} \int_{u=0}^{u=\lambda_2 T} \frac{du}{u^5 \left[\exp\left(\dfrac{C_2}{u}\right) - 1\right]} - \frac{C_1}{\sigma} \int_{u=0}^{u=\lambda_1 T} \frac{du}{u^5 \left[\exp\left(\dfrac{C_2}{u}\right) - 1\right]} \quad [2.88]$$

$$F_{\lambda_1-\lambda_2,T} = G_{\lambda_1 T-\lambda_2 T} = G_{0-\lambda_2 T} - G_{0-\lambda_1 T} \quad [2.89]$$

with:

$$C_1 = 2\pi hc^2 \quad \text{and} \quad C_2 = \frac{hc}{k} \qquad [2.90]$$

This function will be very important in practice. It in fact makes it possible to calculate the emission of a black, or not black, body in a spectral interval using Wien's law, without having to handle the luminance values directly.

It is therefore enough to use a $G_{0-\lambda T}$ table. **This table can be found in the Appendix.**

2.5.2. *An application*

Here, we give a direct and simple application. Applications for these $G_{0-\lambda T}$ functions will be in the much more comprehensive application framework found in **Chapter 5**.

EXAMPLE 2.2.– Visible radiation energy

What is the fraction F of the solar radiation emitted in the visible domain, that is, between $\lambda_1 = 380\ nm$ and $\lambda_2 = 780\ nm$?

The Sun is assimilated to a black body of temperature $T_S = 5777\ K$.

SOLUTION.– Here, we find a typical case of using $G_{0-\lambda T}$ functions. The fraction of the emission from the black body between $\lambda = 0\ nm$ and $\lambda_1 = 380\ nm$ is given by $G_{0-\lambda_1 T}$.

The fraction of the black body emission between $\lambda = 0\ nm$ and $\lambda_2 = 780\ nm$ is given by $G_{0-\lambda_2 T}$.

The fraction of the black body emission between $\lambda_1 = 380\ nm$ and $\lambda_2 = 780\ nm$ is given by $F = G_{0-\lambda_2 T} - G_{0-\lambda_1 T}$.

$\lambda_1 T$ and $\lambda_2 T$ are calculated, remembering that the table supposes that the wavelengths are expressed in microns:

$$\lambda_1 T = 0.38 * 5777 = 2195 \approx 2200 \qquad [2.91]$$

$$\lambda_2 T = 0.78 * 5777 = 4506 \approx 4500 \qquad [2.92]$$

The table gives us:

$$G_{0-2200} = 0.1009 \qquad [2.93]$$

$$G_{0-4500} = 0.5643 \qquad [2.94]$$

The fraction of the black body emission between $\lambda_1 = 380\, nm$ and $\lambda_2 = 780\, nm$ is therefore:

$$F = G_{0-4500} - G_{0-2200} = 0.5643 - 0.1009 = 0.463 \qquad [2.95]$$

The visible spectrum therefore practically represents 46.3% of the emission from the Sun.

We thus discover the importance of Wien's law, which makes it possible to determine the maximum zone of emission. The very strong dynamics of the spectrum of the black body locates an important part of the emission in the region of the maximum of the spatial distribution.

2.6. Emissivity of any body: a general case of a non-black body

2.6.1. *Definition of monochromatic emissivity*

The luminance $L_\lambda(T)$ of an emissive body at a temperature T and a given wavelength λ will be calculated by multiplying the luminance of the black body at the same wavelength and the same temperature $L_\lambda^0(T)$ by a coefficient called monochromatic emissivity, generally written as $\varepsilon_\lambda = \varepsilon(\lambda)$. This coefficient without dimension will evidently be inferior to 1 since the black body by definition is the most energetic emitter. It is generally a function of the wavelength. $\varepsilon_\lambda = \varepsilon(\lambda)$ expresses the result of a knowledge of the material most often acquired experimentally.

$$L_\lambda(T) = \varepsilon_\lambda L_\lambda^0(T) \qquad [2.96]$$

This relationship is fundamental for the examples tackled here. It was written by Kirchhoff in 1859.

In many cases, this emissivity can be considered isotropic; it does not depend on the inclination of the beam in relation to the source. Lambert's law can thus be preserved for any body. In other cases, this incidence should be considered. Monochromatic emissivity may depend on temperature.

In all cases, the radiated energy results from an integral, which is generally calculated using digital calculation.

For a wavelength, there is in fact a given solid angle:

$$d^3\Phi = \varepsilon_\lambda L_\lambda^0(T)\, dS \cos\theta\, d\lambda\, d\Omega \qquad [2.97]$$

The resulting flow on all the wavelengths will be:

$$d^3\Phi = \int_0^\infty \varepsilon_\lambda L_\lambda^0(T)\ dS\cos\theta\ d\Omega\, d\lambda = \left[\int_0^\infty \varepsilon_\lambda L_\lambda^0(T)\ d\lambda\,(\theta)\right] dS\cos\theta\ d\Omega \qquad [2.98]$$

which can be rewritten as:

$$d^3\Phi = \left[\int_0^\infty \varepsilon_\lambda L_{T,\lambda}\ d\lambda\right] dS\cos\theta\ d\Omega \qquad [2.99]$$

A luminance resulting from the body is thus drawn out:

$$L_T = \int_0^\infty \varepsilon_\lambda L_\lambda^0(T)\ d\lambda \qquad [2.100]$$

2.6.2. *Definition of global emissivity: a tricky concept*

In these conditions, a global emissivity ε_G is sometimes defined by:

$$L_T = \int_0^\infty \varepsilon_\lambda L_\lambda^0(T)\, d\lambda = \varepsilon_G L_T^0 \quad [2.101]$$

As well as varying with temperature, **this coefficient is only one means of expressing the result of the complex operation of integration that we have just described**. If used without caution (e.g. by applying the result found for one temperature for another), it will be particularly dangerous. This practice will only be legitimate in a particular instance, pictured below.

2.6.3. *Emissivity of a gray body: a particular case*

a) Defining a gray body

Bearing in mind the previous reflection, we can encounter (and seek wherever possible) a very interesting and particular case.

For some bodies, we admit, sometimes with a degree of approximation, that the emissivity does not vary with the wavelength. Implicitly, it can be seen that this coefficient is also isotropic.

Which is:

$$\varepsilon_\lambda \equiv \varepsilon = Ct \quad [2.102]$$

In this case, the expression of the total luminance is considerably simplified.

We have:

$$d^3\Phi = \varepsilon L_\lambda^0(T)\, dS \cos\theta\, d\lambda\, d\Omega \quad [2.103]$$

therefore:

$$L_T = \int_0^\infty \varepsilon L_\lambda^0(T)\, d\lambda = \varepsilon \int_0^\infty L_\lambda^0(T)\, d\lambda \quad [2.104]$$

We see in passing that the global emissivity ε_G is identified with ε

$$L_T = \int_0^{\infty} \varepsilon L_\lambda^0(T)\, d\lambda = \varepsilon L_T^0 = \varepsilon_G L_T^0 \qquad [2.105]$$

b) Emittance of a gray body

The emittance of a gray body therefore becomes particularly interesting to write.

We know that, by applying Lambert's law:

$$M_\lambda(T) = \pi L_\lambda(T) = \pi \varepsilon L_\lambda^0(T) \qquad [2.106]$$

which can be written as:

$$M_\lambda(T) = \varepsilon M_\lambda^0(T) \qquad [2.107]$$

where we recall that $M_\lambda^0(T)$ is the monochromatic emittance of the black body.

So, for all the spectrum, the relationship between the total emittances is:

$$M(T) = \int_0^{\infty} M_\lambda(T)\, d\lambda = \int_0^{\infty} \varepsilon M_\lambda^0(T)\, d\lambda = \int_0^{\infty} \varepsilon M_\lambda^0(T)\, d\lambda = \varepsilon M^0(T) \qquad [2.108]$$

Indeed, we know from the Stefan–Boltzmann law that:

$$M^0(T) = \sigma T^4 \qquad [2.109]$$

and in consequence:

$$M(T) = \varepsilon \sigma T^4 \qquad [2.110]$$

This relationship, which we recall is valid exclusively for a gray body, will be very useful for us in solving our examples.

c) Gray bodies piecewise

We will encounter the case of a body whose emissivity is not constant, but remains constant piecewise, over finite intervals of a wavelength. Take, for example, the case of a body with an emissivity equal to ε_1 between $\lambda = 0$ and $\lambda = \lambda_1$, then an emissivity ε_2 of $\lambda = \lambda_1$ to $\lambda \to \infty$.

Its total emittance will be calculable in two pieces:

from $\lambda = 0$ *à* $\lambda = \lambda_1$: $M_1 = \pi \int_0^{\lambda_1} L_\lambda(T)\, d\lambda$

$$= \pi \int_0^{\lambda_1} \varepsilon_1 L_\lambda^0(T)\, d\lambda = \pi \varepsilon_1 \int_0^{\lambda_1} L_\lambda^0(T)\, d\lambda \qquad [2.111]$$

from $\lambda = \lambda$ to $\lambda \to \infty$: $M_2 = \pi \int_{\lambda_1}^{\infty_1} L_\lambda(T)\, d\lambda$

$$= \pi \int_{\lambda_1}^{\infty_1} \varepsilon_2 L_\lambda^0(T)\, d\lambda = \pi \varepsilon_2 \int_{\lambda_1}^{\infty_1} L_\lambda^0(T)\, d\lambda \qquad [2.112]$$

We see in $\int_0^{\lambda_1} L_\lambda^0(T)\, d\lambda$ and $\int_{\lambda_1}^{\infty_1} L_\lambda^0(T)\, d\lambda$ the contributions to black body radiation in the two ranges of wavelength considered.

We can calculate:

$$\int_0^{\lambda_1} L_\lambda^0(T)\, d\lambda = G_{0-\lambda_1 T} \int_0^{\infty} L_\lambda^0(T)\, d\lambda \qquad [2.113]$$

$$\int_{\lambda_1}^{\infty} L_\lambda^0(T)\, d\lambda = G_{\lambda_1 T-\infty} \int_0^{\infty} L_\lambda^0(T)\, d\lambda \qquad [2.114]$$

Hence, the contributions to the emittance

$$M_1 = \pi \varepsilon_1 \int_0^{\lambda_1} L_\lambda^0(T)\, d\lambda = \varepsilon_1 G_{0-\lambda_1 T}\, \pi \int_0^{\infty} L_\lambda^0(T)\, d\lambda \qquad [2.115]$$

$$M_2 = \pi \varepsilon_2 \int_{\lambda_1}^{\infty_1} L_\lambda^0(T)\, d\lambda = \varepsilon_2 G_{\lambda_1 T-\infty}\, \pi \int_0^{\infty} L_\lambda^0(T)\, d\lambda \qquad [2.116]$$

And as:

$$\pi \int_0^{\infty} L_{\lambda}^0(T)\, d\lambda = \sigma T^4 \qquad [2.117]$$

$$M = M_1 + M_2 = \varepsilon_1 G_{0-\lambda_1 T}\, \pi \int_0^{\infty} L_{\lambda}^0(T)\, d\lambda + \varepsilon_2 G_{\lambda_1 T-\infty}\, \pi \int_0^{\infty} L_{\lambda}^0(T)\, d\lambda \qquad [2.118]$$

$$M = M_1 + M_2 = \left(\varepsilon_1 G_{0-\lambda_1 T} + \varepsilon_2 G_{\lambda_1 T-\infty}\right) \sigma T^4 \qquad [2.119]$$

2.7. Simple applications

EXAMPLE 2.2.– The surface of a black body emits 3,548 W per m².

1) What is the emittance M_T from this surface?

2) What is its temperature T_W? Give it in Celsius.

SOLUTION.–

1) By defining the emittance:

$$M_{T1} = 3548 W \qquad [2.120]$$

2) The Stefan–Boltzmann law gives us a black body:

$$M_T = \sigma T_W^4 \qquad [2.121]$$

We can easily deduce:

$$T_W^4 = \frac{3548}{5.67.10^{-8}} = 6.26.10^{10}\, K^4 \qquad [2.122]$$

$$T_W = 500.15 K = 227.1°\text{C} \qquad [2.123]$$

EXAMPLE 2.3.– A surface of $S = 43\ m^2$ of a gray body of emissivity $\varepsilon = 0.45$ emits a total flow of $\Phi = 5000\ W$.

1) What is the emittance M_T of this surface?

2) What is its temperature T_W ? Give it in Celsius.

SOLUTION.–

1) By defining the emittance:

$$\Phi = S\, M_T \qquad [2.124]$$

So, the value of this emittance is

$$M_T = \frac{5000}{43} = 116.27\, W.m^{-2} \qquad [2.125]$$

2) The Stefan–Boltzmann law indicates:

$$M_T = \sigma T_W^4 \qquad [2.126]$$

so:

$$T_W^4 = \frac{116,27}{5.67.10^{-8}} = 2.0510^9\, K^4 \qquad [2.127]$$

$$T_W = 212.8\, K = -60.2°\text{C} \qquad [2.128]$$

EXAMPLE 2.4.– Emission from a plate.

A plate painted black can easily be assimilated to a gray body with emissivity $\varepsilon = 0.9$.

It is subject to a radiative thermal flux density $\varphi = 988\, W.m^{-2}$.

This plate, under the influence of strong wind creating a convection coefficient h, maintains its surface temperature at a constant value $T_W = 40\,°\text{C}$.

The ambient air temperature is $T_a = 20\,°\text{C}$.

1) Write the expression of different terms of thermal balance (radiation absorbed and emitted and convection) from the surface of the plate. Suppose that there is no thermal conduction towards the interior of the plate. So, h is the convection coefficient.

2) By deducing the value of the coefficient h which makes it possible to maintain the plate at a constant temperature.

SOLUTION.–

1) Flux density absorbed:

$$\varphi_{abs} = 0.9 * 988 = 889.2 W.m^{-2} \qquad [2.129]$$

Flux density emitted:

$$\varphi_{em} = \varepsilon \sigma T_W^4 = 0.9 * 5.76.10^{-8} * 313^4 = 489.78\ W.m^{-2} \qquad [2.130]$$

The outgoing flow density is deduced by difference. The outgoing flow density will have the expression:

$$\varphi_{conv} = h\left(T_W - T_a\right) = 20\,h \qquad [2.131]$$

2) The thermal balance per surface unit of the plate is written as:

Flow absorbed = Flow emitted + Outgoing flow by convection:

$$\varphi_{abs} = \varphi_{em} + \varphi_{conv} \qquad [2.132]$$

which in digital values is:

$$889.2 = 489.78 + 20\,h \qquad [2.133]$$

$$h = \frac{889.2 - 489.78}{20} \qquad [2.134]$$

$$h = 19.97 W\, m^{-2}.K^{-1} \qquad [2.135]$$

EXAMPLE 2.5.– An object in space

An object sent to space is covered by a refractory material that is an extremely good heat isolator. The surface of this material can be considered a gray body with emissivity $\varepsilon = 0.8$.

This material can tolerate high temperatures, up to 1,000°C. The object is sent to space where it is subject, as a result of its stellar environment, to a radiative flow density $\phi = 10000\ Wm^{-2}$.

1) We are in space, the material conducts very little. So, no transfer via conduction or convection can be envisaged. A thermal equilibrium is established and the surface T_W of the material remains constant.

What, in these conditions, is the expression of the emittance M_T from the surface of the refractory material?

2) What is the temperature of the surface T_W of the refractory material?

3) Will this material "hold"? Or do we need to seek another?

SOLUTION.–

1) All the heat absorbed by radiation, which is $\varepsilon\phi$ per m², should be re-emitted by radiation, which is the emittance

$$M_T = \varepsilon\phi \quad [2.136]$$

2) The emittance of the body is that of a gray body with emissivity ε

$$M_T = \varepsilon\sigma T_W^4 \quad [2.137]$$

$$M_T = \varepsilon\sigma T_W^4 = \varepsilon\phi \quad [2.138]$$

$$0.8 * 5.67.10^{-8} * T_W^4 = 0.8 * 10^4 \quad [2.139]$$

$$T_W^4 = 1.76.10^{11} \quad [2.140]$$

$$T_W = 648\,K = 375°\mathrm{C} \quad [2.141]$$

3) The temperature found $T_W = 648\,K = 375°\mathrm{C}$ is much less than the body volatilization temperature, which will therefore remain.

3

Emission and Absorption

3.1. Introduction

When a radiative flux impacts a surface, all or some of these different simultaneous phenomena are involved: **absorption, reflection and transmission**.

It will be useful to recall here some essential definitions of energy flux:

Emitted flux: Φ_e, quantity of radiant energy (joules) per unit of time; therefore expressed in Watts (W).

Emitted flux density: φ_e, quantity of radiant energy (joules) per unit of time, and per unit of surface; thus it is expressed in Watts per meter squared ($W.m^{-2}$).

Flux received Φ_i : quantity of radiant energy received per unit of time, therefore expressed in Watts (W).

Illumination E_i : quantity of radiant energy received per unit of surface, therefore expressed in $W.m^{-2}$; this is an incident flux density.

3.2. Absorption, reflection, transmission

If a solid surface receives a luminous flux, an illumination is written as E.

Consider a body with a certain thickness exposed to incident radiation. Other heat transfers may be present (conduction, convection, etc.). We are only interested here in the radiative part of the transfers of this body to the outside.

To simplify, suppose that we have a body with plane geometry. Consider a unit of surface area.

The incident radiation is encrypted by radiation E, incident energy flow per surface unit, expressed in $W\,m^{-2}$. This illumination could also be written as an incident flux density φ_i.

It should be noted that, in the most general cases, this energy flow has a spectral distribution: each small wavelength interval $\lambda - \lambda + d\lambda$ contributes to E by a variable input depending on the wavelength, E_λ.

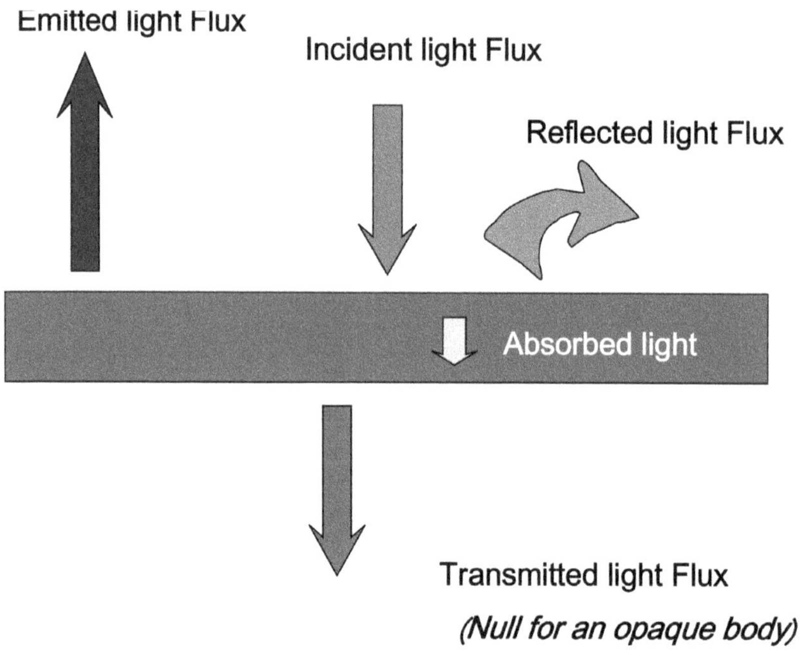

Figure 3.1. *Schematic of radiative exchanges at a surface. For a color version of this figure, see www.iste.co.uk/ledoux/heat2.zip*

E therefore results from an integral:

$$E = \int_0^\infty E_\lambda d\lambda \qquad [3.1]$$

This energy is distributed in three flow densities.

Part of this energy is absorbed by the body. The flux absorbed, φ_{abs}, is:

$$\varphi_{abs} = \alpha_G E \qquad [3.2]$$

Some of this energy is reflected. The flux reflected φ_R is

$$\varphi_R = \rho_G E \qquad [3.3]$$

The reflection coefficient is thus defined as ρ_G, which is a global coefficient.

Some of this energy traverses the body. The flux transmitted Φ_τ is

$$\varphi_\tau = \tau_G E \qquad [3.4]$$

We thus define the transmission coefficient τ_G.

NOTE.– We now stress the fact that coefficients α_G, ρ_G and τ_G are indicator elements, summarizing the results of complex phenomena. In fact, in the most general case, absorption, like transmission and similarly reflection, is dependent on wavelength. We will examine this case in the following chapters. For this reason, we have written them with an index *G*, as global.

These are not in any way intrinsic properties of an illuminated body. Ignoring this observation can lead to serious errors, especially in the examples (for the exercises suggested here, we chose to use simplifying hypotheses, which we will examine later).

These global coefficients are linked by simple equations.

The incident energy is clearly equal to the sum of the energy reflected, transmitted and absorbed.

We therefore have:

$$E = \Phi_R + \Phi_\tau + \Phi_{abs} = \rho_G E + \tau_G E + \alpha_G E = (\rho_G + \tau_G + \alpha_G) E \quad [3.5]$$

So: $\rho_G + \tau_G + \alpha_G = 1$ [3.6]

Many bodies transmit no radiation. They are opaque. In this case, $\tau_G \equiv 0$

Then:

$$\rho_G + \alpha_G = 1 \quad [3.7]$$

and:

$$\rho_G E = E - \alpha_G E \quad [3.8]$$

3.3. Kirchhoff's law

We recall that the global coefficients defined above are, at most, a means of expressing a more detailed calculation.

In this section, we are interested in absorption.

Absorption, as transmission often does, depends on wavelength.

We should therefore write that illumination has a spectral distribution. Let E_λ be the contribution to this spectrum for the small spectral interval between λ and $\lambda + d\lambda$. Then, the illumination can be written as above:

$$E = \int_0^\infty E_\lambda d\lambda \quad [3.9]$$

In the most general case, each spectral component of the illumination is absorbed differently. We thus define a spectral absorption coefficient, α_λ, from the part of the flux absorbed $d\varphi_{abs}(\lambda)$ coming from a given illumination, in the small spectral interval between λ and $\lambda + d\lambda$. We write:

$$d\varphi_{abs}(\lambda) = \alpha_\lambda E_\lambda \qquad [3.10]$$

Kirchhoff studied the relationship between absorption and emission from radiating bodies. In 1859, he published a law that will be fundamental to all applications present in this text. He based his reasoning on studying the thermodynamic equilibrium inside an oven. He expressed the result essentially as follows:

> If a body is subject to radiation from a radiating cavity in thermal equilibrium at temperature T. Depending on the value of the absorption coefficient, this body will absorb some of the radiation that is incident to it. To maintain the equilibrium, it must return the energy received in the same direction and at the same frequency, to replace the energy borrowed from the cavity.

We express this as:

$$\alpha_\lambda = \varepsilon_\lambda \qquad [3.11]$$

At a given wavelength, the absorption coefficient is equal to the emissivity.

This is fundamental for understanding what follows.

3.4. Recap on the global absorption coefficient

3.4.1. *General case*

We know that if a body is not gray, then $\alpha_\lambda = \varepsilon_\lambda$, variable with λ.

α_G will be determined both by this variation in the emission coefficient with the wavelength $\alpha(\lambda)$ and the spectral distribution of the illumination $E(\lambda)$:

$$\alpha_G = \frac{\int_0^\infty \varepsilon(\lambda)E(\lambda)}{\int_0^\infty E(\lambda)} = \frac{\int_0^\infty \varepsilon(\lambda)E(\lambda)}{E} \quad [3.12]$$

We can see that α_G is not an intrinsic coefficient of the material, but on the contrary **depends on the spectral distribution of the illumination**. There is in fact an absorption coefficient for each source illuminating a surface.

NOTE.– Working on a plane surface unit is not reductive. On a surface where the illumination would be heterogeneous, the reasoning made here can easily be extended to an elementary surface dS which will automatically have plane geometry.

Then:

$$E\,dS = \left(\Phi_R + \Phi_\tau + \Phi_{abs}\right) dS \quad [3.13]$$

3.4.2. *Case of the gray body*

If the body is gray, its emissivity no longer depends on the wavelength $\varepsilon_\lambda = \varepsilon$

$$\alpha_G = \frac{\int_0^\infty \varepsilon_\lambda E\,d\lambda}{\int_0^\infty E\lambda d\lambda} = \frac{\varepsilon\int_0^\infty E\,d\lambda}{E} = \varepsilon \quad [3.14]$$

This is the only case with the black body ($\varepsilon = 1$) where the (global) absorption coefficient α_G is intrinsic. It is equal to the emissivity ε (constant with the wavelength):

$$\alpha_G = \varepsilon \quad [3.15]$$

3.5. General case: multiple transfers

In the most general case, the body may be subject to other, different transfers: conductive, notably between surfaces that heat is entering or leaving, convective, either cooling (extracting heat from the body) or heating (introducing heat into the body).

In practice, we will often have to write balances per surface unit of a body, subject to transfers of various natures.

These results will be written for a surface, between different flux densities, if we start with an illumination defined by m^2:

– the illumination or incident flux E or φ_i : this incident flux could come from a source or another radiating surface, etc.;

– a radiated flux, φ_R, determined by the temperature of the equilibrium of the body. This temperature is often an element sought when solving a problem;

– an absorbed flux, φ_{abs};

– a conduction flux φ_{cond}, for example, conducting the heat towards another limit of the body;

– a convective flux φ_{conv}, providing or removing heat by forced or natural convection, etc.;

– a transmitted flux φ_τ, if the body is not opaque.

These fluxes will be allocated a positive or negative sign following the direction of transfer to or from the surface. The sign will generally result automatically from writing this flux.

NOTE.– In writing the transfer that follows – written for the surface – we will consider the absorbed flux as being at the surface, even if the phenomenon of absorption is internal to the body.

The general balance equation written for a surface will therefore have the form (taking into account the fact that the conductive and convective fluxes have a sign):

$$E = \varphi_R + \varphi_{Conv} + \varphi_{abs} + \varphi_{cond} + \varphi_\tau \qquad [3.16]$$

In practice, these different fluxes will be written depending on temperatures.

We will distinguish:

– The wall temperature T_W, the temperature of the surface on which the balance is written.

– An ambient temperature is often denoted by T_a or T_e, which will determine convective transfers. This temperature will often also determine the ways of writing the radiation of the ambience in which the surface is immersed.

We take the example of the thermal equilibrium of a room. It is often considered in simple cases that the room behaves like a black body.

NOTE.– There may be exceptions. In some cases, it will need to be considered that this ambience does not radiate like a black body at an ambient temperature (problems with gaseous radiation, for example, or marked influence of walls that are not in thermal equilibrium with the ambient gas, etc.). In this case, the ambience radiates like a black body at a temperature T_R different from T_a. It is then called radiating temperature T_R.

We note that in the thermal balance that we have to write, this ambient radiation is added to the radiation E imposed by an exterior source (sun, lamp, etc.) E is not always present in all problems (see the examples discussed below).

A temperature T_0, a reference temperature making it possible to calculate conduction. A thermal conductibility λ of the material is often involved.

The convective flux will often be written using a convection coefficient. This coefficient will have a constant evaluated value; in some problems, this can be made more sophisticated by calculating this coefficient from the convection formulae.

As an example, a plane opaque plate, behaving as a gray body with emissivity ε, thickness e and thermal conductibility λ, wall temperature T_W subject to an incident radiation E placed in a room at temperature T_a. It is known that $T_a < T_W$. A draft in the room induces a convection coefficient h. T_0, the temperature on the other side of the plate is such that $T_0 < T_W$.

The balance can be written as:

$$\varepsilon E - \varepsilon\sigma T_W^4 + \varepsilon\sigma T_a^4 - h(T_W - T_a) - \lambda\frac{T_W - T_0}{e} = 0 \qquad [3.17]$$

Here, we know that the conductive and convective fluxes correspond to a loss. The corresponding terms should be negative, hence the minus signs.

We will also note the fairly recurrent presence in this type of balance linking the two terms $\varepsilon\sigma\left(T_a^4 - T_W^4\right)$; this term becomes involved when there is total influence. We will address this important subject and focus on these terms in Chapter 10, section 10.2.1.

We note that, in practice, not all the parameters are known and this balance equation makes it possible to determine one, which is often the wall temperature T_W. Many examples follow, which will enable the reader to become familiar with current calculating practices.

3.6. Absorption: the Beer–Lambert law

3.6.1. *Radiation transfer*

Radiative transfer (or **transfer by radiation**) is the domain of mathematical physics, describing the interaction of electromagnetic radiation and matter. It brings different phenomena into play: absorption, diffusion. It is important but very difficult to address this subject and this is out of the scope of this book. However, in this particularly complex domain, we think it necessary to provide some concepts limited to the phenomenon of absorption, in order to enhance the reader's knowledge.

Absorption consists of a thermalization of radiation, that is, a transfer of energy from an electromagnetic wave in molecular agitation (or atomic in a network).

In a thermal balance, this absorption is counted as a loss, conveyed by a φ_{abs}. In reality, this loss may be due to another cause, which is the diffusion of light: some of the radiation is sent in all directions, often with a change of wavelength. This is due to quantum effects in the atomic layers of the bodies traversed. This phenomenon will not be considered in this book.

Regarding the study of diffusion, we will direct readers who are interested in specialist works or publications. We note that treating the problem fully involves mathematical tools that are difficult for the uninitiated to access.

3.6.2. *Beer's law*

The Beer–Lambert law or **Beer–Lambert–Bouguer law** (or simply, **Beer's law**), is an empirical relationship that links the attenuation of radiation to the chemical properties of the medium it crosses and the thickness traversed. Works on this subject by Bouguer date from 1729.

The Beer–Lambert law establishes a proportionality between the concentration of a chemical entity and the length of the route traveled by the light in the medium considered.

The absorbance A is defined as the ratio between the incident luminance $L_\lambda(0)$ and the luminance measured after a route of length l in the medium $L_\lambda(l)$:

$$A = Ln \frac{L_\lambda(0)}{L_\lambda(l)} \qquad [3.18]$$

The law is expressed in the form:

$$A = K\, l\, C$$

– K is the attenuation coefficient or absorbance coefficient belonging to the medium traversed;

– l is the length of the medium traversed;

– C is the concentration of absorbent molecules in the medium.

Dosimetry techniques may result from these formulas.

The Beer–Lambert law is applied to the absorption of all radiation (electromagnetic, i.e. photons, as well as, by analogy with neutron wave particles, α particles, etc.) by matter and is a basic solution of the radiative transfer equation.

NOTE.– Beer's law does not consider the phenomena of diffusion, which moreover involve reemission of energy in space out of the incident direction, and often give rise to reemission in a wavelength spectrum distinct from that of the incident light.

To model the Beer–Lambert law, we write that the fraction of the energy in the medium considered is independent of this energy and proportional to the length of the path (i.e. to the number of absorbent particles encountered). When defined in terms of luminance, this gives:

$$d\,L_{\lambda,T} = -\,k\;L_{\lambda,T}\,dx \qquad [3.19]$$

where k is the absorption coefficient of the medium traversed.

This transfer equation is integrated easily. If originally, $x=0$, the luminance is $L_{\lambda,T} = L_{\lambda,T}(0)$ (not to be confused of course with $L^0_{\lambda,T}$ which would be the luminance of a black body), the luminance after traversing the medium of length l results from solving:

$$\frac{d\,L_{\lambda,T}}{L_{\lambda,T}} = -\,k\;dx \qquad [3.20]$$

$$Ln\,L_{\lambda,T} = -\,k\,x + Ln\,Cte \qquad [3.21]$$

Applying the original condition:

$$Ln\,L_{\lambda,T}(0) = Ln\,Cte \qquad [3.22]$$

$$Ln\,L_{\lambda,T} = -\,k\,x + Ln\,L_{\lambda,T}\left(0\right) \qquad [3.23]$$

Taking the exponential of each term:

$$L_{\lambda,T} = L_{\lambda,T}\left(0\right)\exp - k\,x \qquad [3.24]$$

which expresses the Beer–Lambert law.

The absorption coefficient is linked via absorbance to the concentration, which is the number N of absorbent particles (absorbers) per unit of volume.

From this perspective, k can be rewritten in different forms:

– in terms of a cross section of the absorbers:

$$k = N\,\sigma \qquad [3.25]$$

where σ is this cross section (of course, expressed in m^2 it has nothing to do with the Stefan constant), which can make it possible to predict a value.

In the case of gaseous media, the number of absorbers per unit of volume, thus by length traversed on a given beam section, is proportional to the density ρ, where, which is equivalent, at a given absolute temperature T is proportional to the pressure through:

$$\frac{p}{\rho} = R\,T \qquad [3.26]$$

The opacity is thus defined χ_{op} such that:

$$k = \chi_{op}\,\rho \qquad [3.27]$$

The literature also contains diagrams making it possible to calculate absorption by gaseous media.

4

Radiation Exchanges Between Surfaces

4.1. Introduction

In a physical system, a surface is able to receive and emit. Incident radiations can result from a radiant energy density contained in a volume containing the surface, for example, in the case of an oven, or simply a room. We will see some examples. This radiation can also come from one or several other identified surfaces. The general problem can be particularly complex.

4.2. Classification

Two situations will be distinguished here:

a) The case of a system of parallel plane surfaces in "total influence". In this case, **both surfaces are considered to be infinitely large** (we call these large in relation to the distance between them); any radiation resulting from one of the surfaces will impact the other. This case will be the most important for all the examples tackled in this manual.

b) The case of systems where the radiating surfaces are of finite size and are not necessarily parallel with one another. Systems with two surfaces can be considered; one example will be tackled here: the number of surfaces may be significant and all may have complex geometry. This is what will be seen, for example, in modeling ovens, which will rapidly lead to a digital approach.

4.3. The case of total influence

This section establishes expressions that are extremely useful for establishing thermal balances for a surface. Many examples of application will be found in **Chapter 5.**

4.3.1. *The case of two parallel plates. Lambert's law*

In total influence, all radiation issuing from one surface impacts the other, in one place or another. This is generally achieved by two parallel plates of indefinite length (indefinite for the modeling)

What will arise from this radiation will of course depend on the emissivity ε from each plate. For a black body plate, all incident radiation will be absorbed; for a gray body, a part will be reflected and re-sent to the emitting plate.

Our objective here is to establish a method for calculating what will generally be the radiative part of a more complete balance. We will therefore focus on one element of a particular plate and write the terms of absorption and emission corresponding to the physical situation. The result will be one element of a more complete balance. The examples that follow will illustrate this step clearly. We now indicate that the reader should approach the topic by comprehending what is happening and not by seeking readymade formulae. We will return to this remark in good time. It is very important.

In the order of complexity, we will consider the system formed by two parallel plates in relation to one another, and of a large size (theoretically infinite) in relation to the distance that separates them. These two plates will be of two different temperatures, T_W and T_a (this choice of notation is made from the perspective of the applications suggested as examples).

We will consider the plate of temperature T_W, and on each plate, we will focus on a small surface dS chosen moreover without particular criteria. What we write will be valid for any surface element.

We will consider three cases:

a) the two surfaces with temperature T_W and T_a are black bodies;

b) the surface with temperature T_W is a gray body of emissivity ε_W, the surface with temperature T_a is a black body;

c) the two surfaces with temperatures T_W and T_a are gray bodies with emissivities ε_W and ε_a, respectively.

All the surfaces will respond to Lambert's law.

In all three cases, we can make two important remarks:

NOTE 4.1.– The total emittance M_W from the plate that interests us will take the form T_W :

$$M_W = \varepsilon_W \sigma T_W^4 \qquad [4.1]$$

with $\varepsilon_W = 1$ in the particular case of a black body.

NOTE 4.2.– The element dS of the surface "*w*" receives, in total influence, radiations from all the points of the surface "*a*".

We will not demonstrate this, but we will note that the **energy received by any element** dS **of the surface "*w*" is calculated formally in the same way in which we establish the relationship giving the emittance from the plate "*w*".**

We will see here that the illumination (energy flux per surface unit) from the surface dS of the surface "*w*" by the surface "*a*" is given by:

$$E = \varepsilon_a \sigma T_a^4 \qquad [4.2]$$

with $\varepsilon_a = 1$ in the particular case of a black body.

4.3.2. *Total influence between two black body surfaces, of temperatures T_W and T_a*

Let us make the "balance" of the energy exiting an element dS of the surface "w":

The light emitted by dS:

$$d\Phi_W = \sigma T_W^4 \, dS \qquad [4.3]$$

The light received and absorbed entirely by dS:

$$d\Phi_a = \sigma T_a^4 \, dS \qquad [4.4]$$

We should stop there; in fact beyond this, both of these terms will be applied in the full balance (which we will do in the examples, including the other conductive or "internal heat generation" terms (electric convector, etc.)).

For simplicity, a "balance of radiative exchange" is often written; from the perspective of the emission, a plus sign is allocated to the black body emission:

$$d\Phi_W = \sigma T_W^4 \, dS - \sigma T_a^4 \, dS = \sigma\left(T_W^4 - T_a^4\right) dS \qquad [4.5]$$

Once we are "in the habit", this writing is instinctive, fast and direct. It can also be extremely dangerous if we do not bear in mind that it represents the sum of two terms of radically different origin: an emission in all directions and an absorption from the plate opposite, in all directions.

4.3.3. *Total influence between two surfaces*

In this section, we study the total influence between two surfaces, one is a gray body with emissivity ε_W and temperature T_W, the other is a black body with temperature T_a.

The rationale will be the same as before; however, one important difference lies in the fact that the body "w" does not absorb everything. The radiative balance will only involve what is emitted and what is absorbed:

– The light emitted by dS:

$$d\Phi_W = \varepsilon_W \sigma T_W^4 \, dS \qquad [4.6]$$

– The light received by dS:

$$d\Phi_a = \sigma T_a^4 \, dS \qquad [4.7]$$

When we make a "retrospective" balance over the surface of "*w*" with the conductive terms, etc., this reflected light will evidently not be counted. The light absorbed reheats the body "*w*", while the light emitted (gray body radiation) cools it. The light reflected $d\Phi_{a\mathrm{Re}f} = (1-\varepsilon_W)\sigma T_a^4 \, dS$ arrives and departs, and has no effect on the balance of "*w*".

So, only the following will be counted:

$$d\Phi_{aW} = \varepsilon_W \sigma T_a^4 \, dS \qquad [4.8]$$

In order to make the same remark as above, for simplicity, it is possible to unite the two components from the radiative part of a balance:

$$d\Phi = \varepsilon_W \sigma T_W^4 \, dS - \varepsilon_W \sigma T_a^4 \, dS = \varepsilon_W \sigma \left(T_W^4 - T_a^4\right) dS \qquad [4.9]$$

4.3.4. *Total influence between two surfaces*

We present the total influence between two surfaces, gray bodies with respective emissivities ε_W and ε_a, and respective temperatures T_W and T_a. The situation is much more complex here. Several calculation methods are needed to arrive at this balance. We will analyze the total radiation that comes from each total surface:

To simplify the writing, since the fluxes are constant across the whole of each surface, we will work using the flux density (flux per surface unit).

We thus define a "flux per m^2" of radiation coming from "*w*", "Φ_W" and a flux coming from "*a*", "Φ_a".

We note that occasionally, we write a flux density using an upper case "Φ".

We write that the heat that leaves per m^2 of surface "*w*", which is φ_W, is formed of the emission from "*w*", which is

$$\varphi_W = \varepsilon_W \sigma T_W^4 \quad [4.10]$$

From what is reflected, from what is received from the surface "*a*", which is

$$(1-\varepsilon_W)\Phi_a \quad [4.11]$$

we write that the heat that leaves per m² of surface "*w*" is Φ_W formed of the emission from "*w*", which is

$$\varphi_W = \varepsilon_W \sigma T_W^4 \quad [4.12]$$

And from what is reflected from what leaves the surface "*w*", which is

$$(1-\varepsilon_W)\Phi_a \quad [4.13]$$

we write that the heat that leaves per m² of the surface "*a*", which is Φ_a that is formed of the emission from "*a*", which is

$$\varphi_a = \varepsilon_a \sigma T_a^4 \quad [4.14]$$

And from what is reflected from what leaves the surface "*a*", which is

$$(1-\varepsilon_a)\Phi_W \quad [4.15]$$

which leads to a system with two equations with two unknowns, Φ_W and Φ_a:

$$\Phi_W = \varphi_W + (1-\varepsilon_W)\Phi_a \tag{4.16}$$

$$\Phi_a = \varphi_a + (1-\varepsilon_a)\Phi_W \tag{4.17}$$

We note that $\varphi_W = \varepsilon_W \sigma T_W^4$ and $\varphi_a = \varepsilon_a \sigma T_a^4$, emissions from two surfaces, are known.

The solution is:

$$\Phi_W = \frac{\varphi_W + (\varepsilon_W - 1)\varphi_a}{1-(1-\varepsilon_W)(1-\varepsilon_a)} \tag{4.18}$$

$$\Phi_a = \frac{\varphi_a + (\varepsilon_a - 1)\varphi_W}{1-(1-\varepsilon_W)(1-\varepsilon_a)} \tag{4.19}$$

No more remains than to make the difference between what arrives at Φ_W and what leaves it, in order to obtain the balance sought:

$$\Phi_W - \Phi_a = \frac{\varepsilon_W \varepsilon_a \sigma (T_W^4 - T_a^4)}{1-(1-\varepsilon_W)(1-\varepsilon_a)} \tag{4.20}$$

4.3.5. *Wall in total influence in an enclosure*

It has already been emphasized that an oven, an enclosed space in thermic equilibrium, is the prototype of a black body. It is even an experimental black body source model. A plate placed in such an oven will receive a black body radiation and the two previous situations can be found, a black or a gray body plate placed in total influence facing a black body, with the same results for the radiative balance.

The same reasoning can be adopted for a simple plate in a room at a constant temperature T_a.

4.3.6. *Important note on the "thermal balance"*

We will see, for example, the case of a convector whose thermal balance will be established between radiated heat, heat absorbed from the "black body" of ambient temperature T_a and the heat provided by the power supply, with heat degraded by the Joule effect. This enables us to know the surface temperature of the convector, for example. We will see this in the examples that follow.

We simply note here that the temperature T_a itself depends on the energy provided by the convector in the room. T_a therefore results from a balance between the energy provided by the Joule effect and the thermal losses from the room. The difference in scale makes it possible to uncouple the two problems completely, examples of which can be tackled in two different chapters of this book.

4.3.7. *A practical approximation*

We have highlighted the term $T_W^4 - T_a^4$ that results from the balance between the radiation emitted and absorbed by a gray (or black) body.

4.3.7.1. *Approximations*

Here, we will demonstrate and discuss an approximation of this term by the expression:

$$T_W^4 - T_a^4 \approx 4T_a^3\left(T_W - T_a\right) \quad [4.21]$$

This approximation is only valid if $T_W - T_a << T_a$. Its uncontrolled application can therefore be dangerous.

The use of this approximation essentially lies in a calculation trick.

Surface heat balances show convection terms, indeed conduction terms, which are linear in $T_W - T_a$, of type $h\left(T_W - T_a\right)$.

If it is known, which is usually the case, the above approximation linearizes a difference in terms of the power 4.

$$\text{In } T_W^4 - T_a^4 \approx 4T_a^3 (T_W - T_a), \quad [4.22]$$

$4T_a^3$ plays the role of a convection coefficient.

This makes it possible to solve a fourth-degree equation by linearizing it, which can be practical when (this happens) we have a calculator without a solver.

4.3.7.2. *Orders of size*

The demonstration that follows is based on an often-encountered mathematical approximation:

$$(1+\varepsilon)^n \approx 1 + n\varepsilon \quad [4.23]$$

which implies that ε is "infinitely" small.

This neat formula results from a limited development. "Infinitely small" is a term that may be vague in physics. It is not a bad idea to test it digitally by giving ε increasing values. Here, we will do this for $n=2$ and $n=4$, successively giving ε the values ε, ε and ε. In Table 10.1, we compare the true values with the "approximate" values.

		$(1+\varepsilon)^n$	$1+n\varepsilon$	Relative error
$n=2$				
$\varepsilon=10^{-3}$		1.002001	1.002	10^{-6}
$\varepsilon=10^{-2}$		1.0201	1.02	$9.8.10^{-5}$
$\varepsilon=10^{-1}$		1.21	1.2	$8.3.10^{-3}$
$n=4$				
$\varepsilon=10^{-3}$		1.004006	1.004	6.10^{-6}
$\varepsilon=10^{-2}$		1.0406	1.04	6.10^{-4}
$\varepsilon=10^{-1}$		1.4641	1.4	$4.38.10^{-2}$

Table 4.1. *Comparison of* $(1+\varepsilon)^n$ *and of its approximate expression* $1+n\varepsilon$

We clearly see that for $\varepsilon = 10^{-2}$, the approximation is still excellent and that at $\varepsilon = 10^{-1}$, the error becomes appreciable, especially for $n = 4$.

4.3.7.3. *Demonstration*

Once these orders of size are acquired, we demonstrate our relationship.

$$T_W - T_a << T_a \quad [4.24]$$

We rewrite:

$$T_W^4 - T_a^4 = -T_a^4 \left[1 - \left(\frac{T_W}{T_a} \right)^4 \right] \quad [4.25]$$

We note that:

$$T_W^4 - T_a^4 = -T_a^4 \left[1 - \left(\frac{T_W}{T_a} \right)^4 \right] \quad [4.26]$$

We can also write:

$$1 - \left(\frac{T_W}{T_a} \right)^4 = 1 - \left(\frac{T_W - T_a}{T_a} + 1 \right)^4 \quad [4.27]$$

As for $T_W - T_a << T_a$, the ratio

$$\varepsilon = \frac{T_W - T_a}{T_a} \text{ is very small } \varepsilon << 1 \quad [4.28]$$

and:

$$\left[\frac{T_W - T_a}{T_a} + 1 \right]^4 = (1+\varepsilon)^4 \approx 1 + 4\varepsilon = 1 + 4\frac{T_W - T_a}{T_a} \quad [4.29]$$

We will finally have:

$$1-\left(\frac{T_W}{T_a}\right)^4=1-\left(1+4\frac{T_W-T_a}{T_a}\right)=-4\frac{T_W-T_a}{T_a} \qquad [4.30]$$

Taking up the original term:

$$T_W^4-T_a^4=-T_a^4\left[1-\left(\frac{T_W}{T_a}\right)^4\right]\approx-T_a^4\left[-4\frac{T_W-T_a}{T_a}\right] \qquad [4.31]$$

$$T_W^4-T_a^4\approx-T_a^4\left[-4\frac{T_W-T_a}{T_a}\right]=4T_a^4\frac{T_W-T_a}{T_a} \qquad [4.32]$$

which is finally the expected relationship:

$$T_W^4-T_a^4\approx 4T_a^3\frac{T_W-T_a}{T_a} \qquad [4.33]$$

For:

$$\varepsilon=\frac{T_W-T_a}{T_a}<<1 \qquad [4.34]$$

4.3.8. *Complex system of radiant finished surfaces: geometric form factor*

We consider, for example, a system of opposing surfaces such as the already complicated system represented in the figure. Each of the nine surfaces emits and exchanges radiation with the eight others.

We allocate each surface an index, which is S_i. The surface S_i is at temperature T_i **and receives** an illumination E_i **from the rest of the system. It has an emissivity** ε_i.

We call J_i**, the sum of the flux densities that leave** S_i (J_i is therefore expressed in Watt per m²).

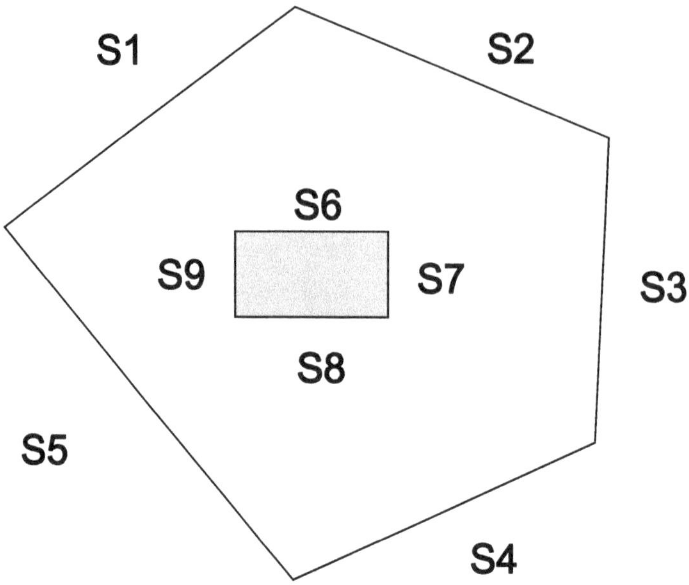

Figure 4.1. *Example of radiation between multiple surfaces*

J_i is formed from an *emitted thermal flux* (gray body) to which the *reflected flux* (illumination received minus the absorbed flux) is added.

Wien's law gives the emitted thermal flux (gray body):

$$\Phi_i = S_i \varepsilon_i \sigma T_i^4 \tag{4.35}$$

The reflected flux = incident flux - absorbed flux = $S_i \left(1 - \varepsilon_i\right) E_i$. [4.36]

This gives:

$$J_i = S_i \varepsilon_i \sigma T_i^4 + S_i \left(1 - \varepsilon_i\right) E_i \tag{4.37}$$

This flux J_i is an emittance from a body obeying Bouguer–Lambert's law. It is therefore allocated a luminance $\dfrac{J_i}{\pi}$.

The exchange between two surfaces can be written as "i" and "j" (see section 1.3.3 for Bouguer–Lambert's law):

$$d^2\Phi_{i\to j} = \frac{J_i}{\pi}\frac{\cos\theta_i \cos\theta_j \, dS_i \, dS_j}{r^2} \quad [4.38]$$

Either by integrating them over all the S_i and all the S_j:

$$d^2\Phi_{i\to j} = \int_{Si}\int_{Sj}\frac{J_i}{\pi}\frac{\cos\theta_i \cos\theta_j \, dS_i \, dS_j}{r^2} \quad [4.39]$$

We can draw from this integral:

$$d^2\Phi_{i\to j} = \frac{J_i}{\pi}\int_{Si}\int_{Sj}\frac{\cos\theta_i \cos\theta_j \, dS_i \, dS_j}{r^2} \quad [4.40]$$

which is:

$$d^2\Phi_{i\to j} = \frac{J_i}{\pi} S_i \, F_{i\to j} \quad [4.41]$$

The geometric form factor has thus been defined $F_{i\to j}$ by:

$$F_{i\to j} = \frac{1}{S_i}\int_{Si}\int_{Sj}\frac{\cos\theta_i \cos\theta_j \, dS_i \, dS_j}{r^2} \quad [4.42]$$

The symmetry of the definition in "i" and "j" causes a symmetry of $F_{i\to j}$:

$$F_{j\to i} = \frac{1}{S_j}\int_{Si}\int_{Sj}\frac{\cos\theta_i \cos\theta_j \, dS_i \, dS_j}{r^2} \quad [4.43]$$

which is the important relationship:

$$S_i \, F_{i\to j} = S_j \, F_{j\to i} \quad [4.44]$$

We can therefore calculate the incident fluxes on each surface. The incident flux Φ_i on S_i is:

$$\Phi_i = S_i\, E_i = \sum_{j=1}^{j=n} \Phi_{j\to i} \tag{4.45}$$

$$\Phi_i = \sum_{j=1}^{j=n} \frac{J_i}{\pi} S_j\, F_{j\to i} \tag{4.46}$$

$$\Phi_i = \sum_{j=1}^{j=n} \frac{J_i}{\pi} S_i\, F_{i\to j} \tag{4.47}$$

By expressing J_i:

$$J_i = \varepsilon_i\, \sigma T_i^4 + \left(1-\varepsilon_i\right) \sum_{j=1}^{4\,j=n} \frac{J_i}{\pi} S_i\, F_{i\to j} \tag{4.48}$$

$$J_i = \sum_{j=1}^{j=n} J_j\, \delta_{ij} \tag{4.49}$$

where δ_{ij} is the Kronecker symbol:

$$\delta_{ij} = 0 \text{ if } i \neq j \tag{4.50}$$

$$\delta_{ij} = 1 \text{ if } i = j \tag{4.51}$$

Finally, a linear system remains:

$$\sum_{j=1}^{j=n} \left[\delta_{ij} - \left(1-\varepsilon_i\right) F_{i\to j} \right] \frac{J_j}{\pi} = \varepsilon_i\, \sigma T_i^4 \tag{4.52}$$

Since the T_is are known, there are n linear equations for the n unknowns that are J_i.

One difficulty will lie in determining form factors. For many geometries, they are known and expressible in the form of algebraic expressions. Suffice to say that the process of solving this system will call on digital methods. This type of calculation is most often required by professionals calculating thermal chambers and ovens. Moreover, this work is designed to be an encyclopedia. We will therefore refer the reader to dedicated works on thermics for the tabulation of these relatively complex expressions. We will, however, give an example of this type of calculation on a very simple case of two surfaces at right angles.

4.3.9. *Application*

EXAMPLE 4.1.– Cooking using a metal plate and aesthetics

A metal cooking plate is a steel heating plate of length $L = 600\,mm$ and size $l = 400\,mm$; the surface has an emissivity $\varepsilon = 0,4$. A thermal resistance of power $P = 3500W$ delivers a uniform flux Φ exclusively via the upper part of the cooking plate. The ambient temperature is $T_a = 21°\text{C}$. A moderate current of air passes over the cooking plate, determining a convection coefficient that will be evaluated at $h = 10\ W.m^{-1}K^{-1}$.

For practical reasons (food spitting fat), we wish to place a decorative vertical plate, perpendicular to the cooking surface. This plate will have the same length and size as the cooking plate. To simplify, suppose that the decorative plate and the heating surface form a dihedral.

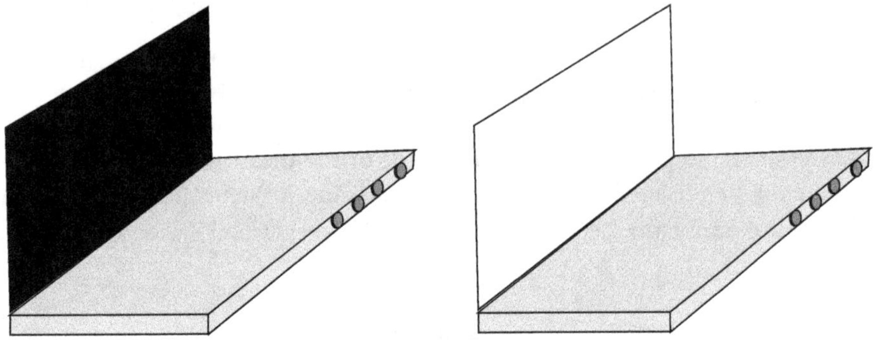

Figure 4.2. *The two plate models*

In such a configuration, we see that the geometric radiation factor has the value

$$F_{1\to2} = F_{2\to1} = 1 - \sin\frac{\theta}{2}$$

where θ is the angle of the two walls.

For aesthetic reasons, the user wishes to paint the vertical plate black, which will give an emissivity $\varepsilon_N = 0.9$. The wall temperature of this plate will be sought here. Intuition also suggests we make this calculation if the plate is painted white, with emissivity $\varepsilon_B = 0.1$.

To simplify the calculations, suppose that the vertical plate is perfectly isolated thermally on its rear side.

1) Calculate the surface temperature T_W of the cooking plate:

a) supposing that there is no air current;

b) taking account of an air current.

At this level, suppose that the decorative plate is not in place.

2) Radiated flux

a) In the two previous cases (questions **1a** and **1b**), what is the flux Φ_{wr} radiated by the cooking plate?

b) What then, in each of these cases, is the radiative flux received from the cooking plate by the decorative plate?

3) We seek to predict the temperature of the decorative plate for each of the colors envisaged. In these calculations, the influence of the radiation Φ_{wd} emitted by the decorative plate on the surface of the cooking plate will be left out.

a) Ignore the action of the air current. What are the respective surface temperatures T_{WN} **and** T_{WB} of the plate painted black and the plate painted white?

b) Consider the action of the air current. What are the surface temperatures of the plate painted black and the plate painted white?

c) The influence of the radiation emitted by the decorative plate has been neglected. Is this hypothesis justified?

d) What color(s) is/are permitted for the decorative plate?

SOLUTION.– It is always useful to make some preliminary calculations.

As in any problem with radiation, absolute temperatures must be used systematically.

The ambient temperature is therefore expressed by

$$T_a = 21°\text{C} = 21 + 273 = 294\,K. \quad [4.53]$$

The surface of the cooking plate is

$$S = 0.6 * 0.4 = 0.24\,m^2. \quad [4.54]$$

To establish the thermal balances, it will be more practical to think in terms of flux density

φ_W provided by the electrical resistance:

$$\varphi_W = \frac{P}{S} = \frac{3500}{0.24} = 14583\,W.m^{-2} \quad [4.55]$$

The angle between the two radiating walls is $\theta = \frac{\pi}{2}$. The geometric radiation factor has the value:

$$F_{1\to2} = F_{2\to1} = 1 - \sin\frac{\pi}{4} = 1 - 0.707 = 0.293 \quad [4.56]$$

1) Calculating the surface temperature T_W of the cooking plate

The surface temperature will be determined by a thermal balance "at the surface" of the cooking plate. All of the "incident" fluxes bringing heat to the surface will be counted positively. The fluxes "exiting" the surface will be counted negatively. Since the surface temperature is constant, these fluxes should balance one another. This balance will be established for each surface unit of the cooking plate.

The decorative plate is not put in place; therefore, it is assumed that no flux will be transmitted from this plate to the cooking plate.

a) *If there is no air current*, the balance of fluxes at the surface will become involved:

– the "incident" flux coming from the electrical resistance, expressed in flux density φ_W ;

– the "incident" flux absorbed from the radiation in the atmosphere assimilated, as usual, into a black body, in flux density $\varepsilon \sigma T_a^4$;

– the flux "exiting" from the radiation from the wall, in flux density $\varepsilon \sigma T_w^4$.

The balance is thus written as:

$$\varphi_W + \varepsilon \sigma T_a^4 = \varepsilon \sigma T_w^4 \quad [4.57]$$

which is a fourth-degree equation in T_w, making an explicit solution possible:

$$T_w^4 = \frac{\varphi_W + \varepsilon \sigma T_a^4}{\varepsilon \sigma} \quad [4.58]$$

$$T_w^4 = \frac{14583 + 0.4 * 5.67.10^{-8}\, 294^4}{0.4 * 5.67.10^{-8}} = \frac{1.475.10^4}{2.26.10^{-8}} \quad [4.59]$$

$$T_w^4 = 6.50.10^{11} \quad [4.60]$$

$$T_w = 898\, K = 625\,°\text{C} \quad [4.61]$$

NOTE 4.3.– From this, we can attempt to find the color of the cooking plate using Wien's law:

$$\lambda_{max} T = 2898.10^{-6}\, m.K = 2898\, \mu m.K \qquad [4.62]$$

$$\lambda_{max} = \frac{2898\, \mu m.K}{T_w} = 3.22\, \mu m \qquad [4.63]$$

which clearly lies in infrared.

b) If there is an air current, the balance of the fluxes at the surface involves the same terms as before, by adding to them the "exiting" flux coming from convection:

– the "incident" flux coming from the electrical resistance, flux density φ_W;

– the "incident" flux absorbed coming from the radiation in the atmosphere assimilated, as usual, to a black body, flux density $\varepsilon \sigma T_a^4$;

– the flux "exiting" from the radiation from the wall, flux density $\varepsilon \sigma T_w^4$;

– the flux "exiting" resulting from the convection, flux density

$$h(T_W - T_a). \qquad [4.64]$$

The balance is thus written as:

$$\varphi_W + \varepsilon \sigma T_a^4 = \varepsilon \sigma T_w^4 + h(T_w - T_a) \qquad [4.65]$$

which is a fourth-degree equation in T_W, which will have to be solved numerically t

$$\varepsilon \sigma T_w^4 + hT_w - hT_a - \varphi_W - \varepsilon \sigma T_a^4 = 0 \qquad [4.66]$$

$$0.4 * 5.67.10^{-8} * T_w^4 + 10T_w - 10 * 294 - 14583 - 0.4 * 5.77.10^{-8} * 294^4 = 0 \qquad [4.67]$$

$$2.268.10^{-8}\, T_w^4 + 10T_w - 2940 - 14583 - 169.44 = 0 \qquad [4.68]$$

$$2.268.10^{-8}\, T_w^4 + 10T_w - 14712 = 0 \qquad [4.69]$$

After using a solver, this results:

$$T_w = 750.7\,K = 477.7°\text{C} \qquad [4.70]$$

2) Radiated flux

a) The flux density radiated by the cooking plate

The flux density radiated by the cooking plate, φ_{wR}, is expressed by:

$$\varphi_{wR} = \varepsilon \sigma T_w^4 \;. \qquad [4.71]$$

Without considering the air current, we would calculate:

$$\varphi_{wR} = \varepsilon \sigma T_w^4 = 14748 W.m^{-2} \qquad [4.72]$$

Φ_{wr} here is higher than the flux delivered by the resistance. In fact, the plate should also re-emit the atmospheric radiation absorbed.

Considering the air current:

$$\varphi_{wR} = \varepsilon \sigma T_w^4 = 0.2 * 5.67 * 10^{-8} * 750.7^4 = 7203\, W \qquad [4.73]$$

b) The radiative flux φ_d received by the decorative plate is a fraction of the flux radiated by the cooking plate, given by the radiation factor.

$$\varphi_d = F_{1\to 2}\, \varphi_{wR} \qquad [4.74]$$

with

$$F_{1\to 2} = 0.293 \qquad [4.75]$$

Without considering the air current, we would have:

$$\varphi_d = 0.293 * 14748 = 4321\, W.m^{-2} \qquad [4.76]$$

Considering the air current:

$$\varphi_d = 0.293 * 7203 = 2110\ W.m^{-2} \qquad [4.77]$$

3) Temperature of the decorative plate

We will write the balances in the same spirit as question 1, not considering the radiation φ_{wd} emitted by the decorative plate on the surface of the cooking plate.

a) Neglecting the air current

i) Plate painted black

The balance of the densities of flux at the surface become involved

– the part absorbed from the "incident" flux coming from the cooking plate, $\varepsilon_N \varphi_d$;

– the "incident" flux coming from the radiation from the atmosphere assimilated, as usual, to a black body, σT_a^4;

– the flux "exiting" from the radiation from the wall, $\varepsilon \sigma T_w^4$.

The balance is thus written as:

$$\varepsilon_N \varphi_d + \varepsilon_N \sigma T_a^4 - \varepsilon_N \sigma T_w^4 = 0 \qquad [4.78]$$

which is a fourth-degree equation in T_w, making an explicit solution possible:

$$T_w^4 = \frac{\varepsilon_N \varphi_d + \varepsilon_N \sigma T_a^4}{\varepsilon_N \sigma} \qquad [4.79]$$

$$T_w^4 = \frac{0.9 * 4321 + 0.9 * 5.67.10^{-8} * 294^4}{0.9 * 5.67.10^{-8}} \qquad [4.80]$$

$$T_w^4 = 8.37.10^{10} \qquad [4.81]$$

$$T_w = 537.8K = 264.8°\text{C} \qquad [4.82]$$

ii) Plate painted black

We are led to this, following the same reasoning (only ε_N is replaced by ε_B):

$$T_w^4 = \frac{\varepsilon_B \varphi_d + \varepsilon_B \sigma T_a^4}{\varepsilon_B \sigma} \quad [4.83]$$

$$T_w^4 = \frac{0.2 * 4321 + 0.2 * 5.67.10^{-8}\, 294^4}{0.2 * 5.67.10^{-8}} \quad [4.84]$$

$$T_w^4 = 8.37.10^{10} \quad [4.85]$$

$$T_w = 537.8K = 264.8°\text{C} \quad [4.86]$$

We find the same result, with the emissivities being simplified in the ratios calculated above

b) Considering the air current

What leaves the plate is equal to what goes in. The equation that needs to be solved becomes:

$$\varepsilon \sigma T_w^4 + h\left(T_w - T\right)_a = \varepsilon \varphi_d + \varepsilon \sigma T_a^4 \quad [4.87]$$

which gives, for the plate painted black

$$\varepsilon_N \sigma T_w^4 + hT_w - hT_a - \varepsilon_N \varphi_d - \varepsilon_N \sigma T_a^4 = 0 \quad [4.88]$$

which is the calculation

$$0.9 * 5.67.10^{-8} * T_w^4 + 10T_w - 10T_a - 0.9 * \varphi_d - 0.9 * 5.67.10^{-8}\, T_a^4 = 0 \quad [4.89]$$

$$5.1.10^{-8} * T_w^4 + 10T_w - 2940 - 0.9 * 2110 - 0.9 * 5.67.10^{-8}\, 294^4 = 0 \quad [4.90]$$

$$5.1.10^{-8} * T_w^4 + 10T_w - 5205 = 0 \quad [4.91]$$

$$T_w = 395.6\, K = 122.6°\text{C} \quad [4.92]$$

And for the plate painted white

$$\varepsilon_B \sigma T_w^4 + hT_w - h T_a - \varepsilon_B \varphi_d - \varepsilon_B \sigma T_a^4 = 0 \qquad [4.93]$$

which is the calculation

$$0.2*5.67.10^{-8}*T_w^4 + 10T_w - 10\,T_a - 0.2*\varphi_d - 0.2*5.67.10^{-8}\,T_a^4 = 0 \qquad [4.94]$$

$$1.134.10^{-8}*T_w^4 + 10T_w - 2940 - 0.2*2110 - 1.134.10^{-8}*293^4 = 0 \qquad [4.95]$$

$$1.134.10^{-8}*T_w^4 + 10T_w - 3462 = 0 \qquad [4.96]$$

$$T_w = 332.3\,K = 59.3°\text{C} \qquad [4.97]$$

c) We have neglected the influence of the radiation emitted by the decorative plate.

Let us encrypt the flux φ_P sent by the decorative plate onto the cooking plate, for example, in the case of the black plate with the air current:

$$\varphi_P = \varepsilon_N \sigma T^4 = 0.9*5.67.10^{-8}*395.6^4 \qquad [4.98]$$

$$\varphi_P = 1242\,W.m^{-2} \qquad [4.99]$$

which represents $8,4\%$ of the radiation from the cooking plate itself. Our calculation is therefore approximate, but remains valid in orders of size.

d) What choice should therefore be made concerning the color of the decorative plate?

This choice will depend heavily on the material from which this plate is made. In the absence of an air current, wood is excluded as a material.

In the presence of an air current, if it is wooden, probably melamine, the color white will already be limited. Black would not be a good idea; at more than 100°C, the cooking plate will not have a long guaranteed lifetime.

For a metal plate, it will be necessary to ensure that the paint is resistant to high temperatures.

The ideal of a reflecting plate will still remain, which is often the case in commercial devices. It is, in fact, pleasant to be able to cook outside when there is no wind!

5

Analytic Applications

5.1. Introduction

In these chapters, you will find applications implementing all the knowledge you have acquired in the previous chapters. Integrated into this is an understanding of radiation in the problems that are given to you to solve, often linked to practical devices or circumstances in everyday life; these most often use several modes of thermal transfers.

5.2. Radiators, convectors and radiating fins

EXAMPLE 5.1.– Study of an electric radiator

To heat a room, we use a radiator whose resistance is a cylinder of diameter $D = 2\,\text{cm}$ and length $L = 0.5\,m$. The resistance of this radiator radiates over a lateral surface such as a black body and emits a power $P = 1\,kW$. We ignore exchanges by convection and conduction.

1) This radiator is tested in a laboratory in a vacuum chamber. Calculate the temperature T_W of the radiator.

2) Determine the wavelength $\lambda_{\max}$ for which the spectral energy density emitted by the radiator is maximum.

3) What should the temperature of the radiator be for this wavelength to be $2\,\mu m$? What will the power output then be?

SOLUTION.–

1) In this problem, there is only radiation without absorption by the surface of the radiator. The surface of the radiator behaves like a black body, so Stefan's law applies. The emittance from the radiator is written as:

$$M^0_{T_w} = \sigma T_w^4 \quad [5.1]$$

with

$$\sigma = 5.67.10^{-8}\, W.m^{-2}.K^{-4} \quad [5.2]$$

This emittance is an energy emitted per unit of time and surface. The power P is the energy emitted per unit of time by the total surface of the radiator. We calculate S:

$$S = \pi\, D\, L = 3.14.10^{-2}\ m^2 \quad [5.3]$$

which is the emittance:

$$M^0_{T_w} = \frac{P}{S} \quad [5.4]$$

$$M^0_{T_w} = \frac{1000}{3.14.10^{-2}} = 3.18.10^4\, W.m^{-2} \quad [5.5]$$

We can deduce T_w:

$$M^0_{T_w} = \sigma T_w^4 \quad [5.6]$$

$$3.18.10^4 = 5.67.10^{-8}\, T_w^4 \quad [5.7]$$

$$T_w = 865\, K = 592°\text{C} \quad [5.8]$$

2) Wien's law gives us:

$$\lambda_{\max} T = 2898 \mu m.K \quad [5.9]$$

$$\lambda_{\max} = \frac{2898}{T_w} = 3.35m \qquad [5.10]$$

3) Using Wien's law again:

$$\lambda_{\max} = 2\mu m \qquad [5.11]$$

$$T_w = \frac{2898}{\lambda_{\max}} = 1449K = 1176°C \qquad [5.12]$$

EXAMPLE 5.2.– Convector mural 1

A "convector" fixed on a wall exchanges heat with a room through a surface $1.6m^2$ and is assumed to be flat. The convector is supplied by a water circuit that maintains its temperature T_W at 50°C.

The ambient temperature T_a of the room heated by the radiator is 20°C. The convection coefficient at the surface of the "convector" is taken, equal to $h = 5W\,m^{-2}\,K^{-1}$. The convector is painted cream; it will be considered to be a gray body with emissivity $\varepsilon = 0.2$.

1) What is the heat flux Φ_{conv} communicated to the room by convection?

2) What is the heat flux Φ_R exchanged with the room by radiation?

3) What is the total flux Φ_T communicated to the room. Does the convector really earn its name?

SOLUTION.–

1) By defining the convection coefficient:

$$\Phi_{conv} = Sh(T_W - T_a) \qquad [5.13]$$

$$\Phi_{conv} = Sh(T_W - T_a) \qquad [5.14]$$

$$\Phi_{conv} = 1.6 * 5 * 30 = 240W \qquad [5.15]$$

2) We have:

$$T_W = 323\,K \qquad [5.16]$$

$$T_a = 293\,K \qquad [5.17]$$

The balance between the fluxes emitted and the flux absorbed is written as:

$$\Phi_R = S\,\varepsilon\,\sigma\left(T_W^4 - T_a^4\right) \qquad [5.18]$$

$$\Phi_R = 1.6 * 0.2 * 5.76.10^{-8} * \left(323^4 - 293^4\right) = 318.8\,W \qquad [5.19]$$

3) The total flux is the sum of the balance of the radiated flux and the convected flux

$$\Phi_T = \varphi_R + \varphi_{con} \qquad [5.20]$$

$$\Phi_T = 240 + 319 = 559\,W \qquad [5.21]$$

Convection only represents 43% of the transfer.

EXAMPLE 5.3.– Wall convector 2

An electric convector with a power of 1,000 W radiated in a room via a surface that is assumed to be flat of $S = 3\,m^2$.

The temperature of the room is $T_a = 21°\text{C}$. We call T_W the still-unknown temperature that the surface of radiator takes. The radiator is considered a gray body with emissivity $\varepsilon = 0.5$.

1) If the flux expression is Φ_R exchanged by the radiation between the radiator and the room

$$\Phi_R = S\,\varepsilon\,\sigma\left(T_W^4 - T_a^4\right) \qquad [5.22]$$

2) Deduce from this the value of T_W.

$$\Phi_R = 1000\,W \qquad [5.23]$$

$$T_a = 294\,K \qquad [5.24]$$

$$\left(T_W^4 - T_a^4\right) = \frac{\Phi_R}{S\,\varepsilon\,\sigma} = \frac{1000}{3*0.5*5.67.10^{-8}} = 1.176.10^{10} \qquad [5.25]$$

$$T_W^4 = 1.176.10^{10} + T_a^4 = 1.176.10^{10} + (21+273)^4 = 1.92.10^{10} \qquad [5.26]$$

$$T_W = 372.4\,K = 99.4\,°\text{C} \qquad [5.27]$$

3) We consider a convection coefficient h equal to $h = 5\,Wm^{-2}\,K^{-1}$.

$$\Phi_{conv} = S\,h\left(T_W - T_a\right) = 3*5*(99.4-21) = 1176\,W \qquad [5.28]$$

The convective flux would exceed the electric power provided to the radiator.

The hypothesis that the whole transfer is radiative is therefore clearly false. We note, in passing, that the temperature calculated for the radiator, close to $100°\,C$, will make a particularly dangerous device in a domestic space!

EXAMPLE 5.4.– Convector 3

A "convector" fixed on a wall exchanges heat with a room through a surface that is supposed to be flat $S = 1.6\,m^2$.

The convector is supplied by a water circuit that maintains its temperature at $T_W = 50°\,C$.

The ambient temperature of the room heated by the radiator is $T_a = 20\,°C$. The convection coefficient at the surface of the "convector" is taken to be equal to $h = 5\,W.m^{-2}.K^{-1}$.

The convector is painted cream; it will be considered as a gray body with emissivity $\varepsilon = 0.2$.

1) What is the heat flux Φ_{conv} communicated to the room by convection?

2) What is the heat flux Φ_R exchanged with the room by radiation?

3) What is the total flux Φ_T communicated to the room. Does the convector really earn its name?

SOLUTION.–

1) By defining the convection coefficient:

$$\Phi_{conv} = S\,h(T_W - T_a) \quad [5.29]$$

$$\Phi_{conv} = 1.6 * 5 * 30 = 240W \quad [5.30]$$

2) In the matter of radiation, temperatures should be expressed in Kelvin:

$$T_W = 50 + 273 = 323\,K \quad [5.31]$$

$$T_a = 20 + 273 = 293\,K \quad [5.32]$$

The flux communicated to the room is the balance between the heat radiated by the radiator, a gray body at temperature T_W, and the radiation emitted by the room, a black body at temperature T_W, absorbed by the gray body, which is the radiator, with an absorption coefficient equal to its emissivity:

$$\Phi_R = S\,\varepsilon\,\sigma T_W^4 - S\,\varepsilon\,\sigma T_a^4 = S\,\varepsilon\,\sigma\left(T_W^4 - T_a^4\right) \quad [5.33]$$

$$\Phi_R = 1.6 * 5.67.10^{-8} * \left(323^4 - 293^4\right) \quad [5.34]$$

$$\Phi_R = 63.8W \quad [5.35]$$

3) The flux communicated to the room is the sum of the convected flux

And the radiated flux

$$\Phi_T = \Phi_{conv} + \Phi_R \quad [5.36]$$

$$\Phi_T = 240 + 63.8 \quad [5.37]$$

$$\Phi_T = 303.8W \quad [5.38]$$

Convection represents only 21% of the transfer.

EXAMPLE 5.5.– Radiating fins

We return to radiating fins, which are tackled in section A.6.2. of Appendix 6 of Volume 1. Here, we consider a transfer component in the form of radiation.

A radiating fin is attached to a tube. The surface of the fin is $S = 40\text{cm}^2$. This tube is at the temperature $T_0 = 75°\text{C}$. The whole radiating fin tube is placed in an ambient atmosphere whose temperature is $T_a = 18\,°\text{C}$.

We recall that the efficacity η of a radiating fin is defined as the ratio of the flux it dissipates to the flux it *would* dissipate if it was isothermic at temperature T_0.

1) Suppose, in the first instance, that the heat transfer between the radiating fin and the exterior occurs through convection. We take a convection coefficient $h = 7.5W.m^{-2}.K^{-1}$. We give the efficiency of the radiating fin, equal to $\eta = 0.9$.

What flux Φ_{conv} is the radiating fin capable of evacuating by convection?

2) In a second instance, we evaluate the heat transfer between the radiating fin and the exterior, which can occur via radiation. In the initial project, the emissivity of material the radiating fin is made of is $\varepsilon_1 = 0.2$. What then would be the flux Φ_{ray1} evacuated by radiation? It will be considered that the efficacity of the radiating fin is identical for convective and radiative transfers.

3) We treat the radiating fin so that its emissivity moves to $\varepsilon_2 = 0.9$. What does the flux evacuated by radiation, which is Φ_{ray2}, become?

We now consider two modes of transfer for the radiating fin. We wish to evacuate a power $P = 2.8W$. Is it necessary to treat the fin?

SOLUTION.–

1) The flux is convective:

If the whole of the radiating fin was T_0, the flux would be:

$$\Phi_{conv} = S\,h(T_0 - T_a) \quad [5.39]$$

Considering the efficacity equal to η, the convective flux will be:

$$\Phi_{conv} = 0.9 * 40.10^{-4} * 7.5 * (75 - 18) \quad [5.40]$$

$$\Phi_{conv} = 0.9 * 40.10^{-4} * 7.5 * (75 - 18) \quad [5.41]$$

which is:

$$\Phi_{conv} = 1.54W \quad [5.42]$$

2) We retain the value of the efficacity, so, in the same way, the radiative flux for an emissivity $\varepsilon_1 = 0.2$ will be:

$$\Phi_{rad1} = \eta\, S\, \varepsilon_1\, \sigma\left(T_0^4 - T_a^4\right) \quad [5.43]$$

which is:

$$\Phi_{rad1} = 0.9 * 40.10^{-4} * 0.2 * 5.67.10^{-8}\left[(273 + 75)^4 - (273 + 18)^4\right] \quad [5.44]$$

$$\Phi_{rad1} = 0.9 * 40.10^{-4} * 0.2 * 5.67.10^{-8}\left[1.467.10^{10} - 7.17.10^{9}\right] \quad [5.45]$$

$$\Phi_{rad1} = 0.596\ W \quad [5.46]$$

NOTE.– At radiative level, we meet an initial difficulty. The efficacity of a radiating fin is generally determined using a convective transfer model. Its value can therefore be modified in the case of a radiative transfer. Here, we keep this efficacity value because it is prescribed for us in the text. We need to be aware of the approximative character of this approach.

3) This question is more difficult that it seemed at first.

In fact, the theory of radiating fins shows that the radiating fin is not isothermic and that its temperature decreases from the support to cool towards the exterior. A radiation calculation assuming a homogeneous temperature is therefore an optimistic one.

In any case, we should return to the theory of the radiating fin by introducing the term radiation. The result would be a nonlinear equation that is difficult to solve analytically.

We should therefore reason in order of size.

In the case of the untreated radiating fin, the sum of the flux, calculated here by convection and radiation, which we recall over-evaluated the flux actually evacuated, would give:

$$\Phi_{conv} + \Phi_{R1} = 1.53 + 0.53 = 2.06W \qquad [5.47]$$

an overestimated result that remains lower than $P = 2.3W$.

The untreated radiating fin will not work.

In the case of the treated radiating fin, the sum of the fluxes calculated here by convection and radiation, which always over-evaluates the flux really evacuated, would give:

$$\Phi_{conv} + \Phi_{R2} = 1.53 + 1.53 = 3.06W \qquad [5.48]$$

This result, although overestimated, is much higher than $P = 2.3W$.

Since the efficiency of the radiating fin is $\eta = 0.9$ we can suppose that the temperature at the free end of the fin would not be much lower than T_0.

In this case, the radiation will not be drastically overestimated and we can believe, with good probability, that the radiating fin, which should, in any case, be treated, will work.

Answer: yes, the fin should be treated.

NOTE.– The reasoning above could seem to be only superficially plausible to some. However, it is mainly a problem for the engineer. In the past, was it better to be satisfied with a "back of the envelope" calculation such as the one above, which is economical and passes tests, with a good probability of success? Or is it better to calculate digitally, which takes lengthier intellectual investment, to increase its probability of success, which does not mean dispensing with tests? We cannot escape the fact that, on the ground, the engineer may be both technician and manager.

EXAMPLE 5.6.– Cooling a transistor

A power transistor should dissipate $\dot{.}$. We attach black radiating fins to it, with a surface $S = 20\,\text{cm}^2$. The convection coefficient is $h = 6\ W.m^{-2}.K^{-1}$). The ambient temperature is $T_a = 20°\,C$. We ignore losses by conduction.

1) Which process, either radiation or convection, is the one that would dissipate only the best energy produced? What would the temperature T_W of the radiating fins be in each of these?

2) Both processes dissipate at the same time. What is the equation making it possible to determine the temperature T_w of the radiating fins? How do we determine the numerical value of T_w?

SOLUTION.–

1) The flux density to be dissipated is:

$$\varphi = \frac{P}{S} \qquad [5.49]$$

$$P = \frac{9*4.18}{60} = 0.627W \qquad [5.50]$$

$$\varphi = \frac{0.627}{20.10^{-4}} = 313.5W.m^{-2} \qquad [5.51]$$

In the case of pure convection, the wall temperature T_{WConv} meets

$$\varphi = h\left(T_w - T_a\right) \qquad [5.52]$$

which is:

$$T_w - T_a = \frac{\varphi}{h} = \frac{313.5}{6} = 52.25 \qquad [5.53]$$

$$T_w = 52.25 + T_a = 72.25°C \qquad [5.54]$$

In the case of pure radiation, the wall temperature T_{Wray} meets

$$\varphi = \varepsilon\sigma\left(T_w^4 - T_a^4\right) \qquad [5.55]$$

with $\varepsilon = 1$, since the radiating fins are black.

$$T_w^4 - T_a^4 = \frac{\varphi}{\sigma} = \frac{313.5}{5.67.10^{-8}} = 5.529.10^9 \qquad [5.56]$$

$$T_w^4 = 5.529.10^9 + \left(273+20\right)^4 = 1.29.10^{10} \qquad [5.57]$$

$$T_w = \left(1.29.10^{10}\right)^{0,25} = 337\,K = 64°C \qquad [5.58]$$

The process that evacuates the power for a minimum wall temperature is the radiative process. **This process is therefore more efficient.**

2) When both processes are present, the expression of the flux density is in the order of:

$$\varphi = h\left(T_w - T_a\right) + \sigma\left(T_w^4 - T_a^4\right) \quad [5.59]$$

At this stage, we can solve this equation using a solver, present as a general rule on scientific calculators.

Failing that, and noting that $T_w - T_a \approx 337 - 293 = 44\,K \approx 0.13T_w \approx 0.15T_a$, we can try to use the relationship:

$$T_w^4 - T_a^4 = 3T_a^3\left(T_w - C\right) \quad [5.60]$$

This relationship with the previous values is verifiable at nearly 40%

$$T_w^4 - T_a^4 = 5.53.10^9 \quad [5.61]$$

$$3T_a^3\left(T_w - T_a\right) = 3 * 2.51.10^7 * 44 = 3.31.10^9 \quad [5.62]$$

The equation becomes:

$$\varphi = h\left(T_w - T_a\right) + 3\sigma T_a^3\left(T_w - T_a\right) \quad [5.63]$$

$$\varphi = h\left(T_w - T_a\right) + 3\sigma\, T_a^3\left(T_w - T\right) = \left(h + 3\sigma T_a^3\right)\left(T_w - T_a\right) \quad [5.64]$$

$$313.5 = \left(5 + 3 * 5.67.10^{-8} * 293^3\right)\left(T_w - T_a\right) = \left(5 + 4.28\right)\left(T_w - T_a\right) \quad [5.65]$$

$$T_w - T_a = \frac{313.5}{9.28} = 33.78°\text{C} \quad [5.66]$$

$$T_w = \frac{313.5}{9.28} = 33.78 + T_a = 53.78°\text{C} \quad [5.67]$$

We verify that we naturally find a temperature lower than both T_{wconv} and T_{wrad}.

EXAMPLE 5.7.– Emission from a plate

A treated plate has an emissivity $\varepsilon = 0.8$ between wavelengths $\lambda_1 = 3\,\mu m$ and $\lambda_2 = 4\,\mu m$. Its emissivity is null outside this wavelength band.

This plate is brought to a temperature of $T = 450°\text{C}$.

1) What is the fraction of the total emission from a black body that is sent between λ_1 and λ_2? If the plate was a black body, what would its emittance M_T^0 be?

2) Considering the real value of its emissivity in the zone where it radiates, what is the emittance M_T from this plate?

3) At what wavelength does this plate emit a maximum radiation?

SOLUTION.–

1) We will calculate the fraction emitted between λ_1 and λ_2 using the function $G_{\lambda_1 T-\lambda_2 T} = G_{0-\lambda_2 T} - G_{0-\lambda_1 T}$

$$T = 723\,K$$

$$\lambda_1\,T = 2169$$

$$\lambda_2\,T = 2892$$

$$G_{2892-2869} = 0.2483 - 0.0954 = 0.153 \quad [5.68]$$

The fraction is

$$f = 15.3\%. \quad [5.69]$$

The emittance from the black body at 723 K is:

$$\sigma\,T^4 = 5.67.10^{-8}\,(723)^4 = 1.549.10^4\,Wm^{-2} \quad [5.70]$$

$$M_T^0 = f\sigma T^4 = 0.153*1.549.10^4 = 2370 Wm^{-2} \quad [5.71]$$

2) If the plate is considered as a gray body with emissivity ε

$$M_T = \varepsilon f \sigma T^4 = 0.8*0.153*1.549.10^4 = 1896\ Wm^{-2} \quad [5.72]$$

3) Wien's first law shows us that, at 723 K, the maximum emission from a black body is λ_m such that

$$\lambda_m T = 2898 \quad [5.73]$$

which is

$$\lambda_m = \frac{2898}{723} = 4.008\,\mu m \quad [5.74]$$

which is in the plate's emission zone. Therefore, this is also the wavelength sought.

EXAMPLE 5.8.– Surface with complex emissivity

A flat plate of 10 m² has a fixed temperature $T_W = 450°\,C$.

It has undergone a surface treatment that leads to a variable emissivity ε_λ with a wavelength following the curve given below. Note that beyond $\lambda = 6\,\mu m$, the emissivity ε_λ is null.

We recall the value of Stefan–Boltzmann's constant: $\sigma = 5.67.10^{-8}\ SI$.

1) Calculate the total emittance M_T from this plate at a given temperature T_W. Deduce from this the flux Φ emitted by this plate.

2) What is the total emissivity ε from the plate in these conditions?

3) What is the maximum emissivity wavelength λ_m of a black body that would be at the temperature of the plate?

4) Can you easily evaluate the wavelength at which the plate emits a maximum radiation? If so, what is it?

5) Independent question: Is the plate a gray body?

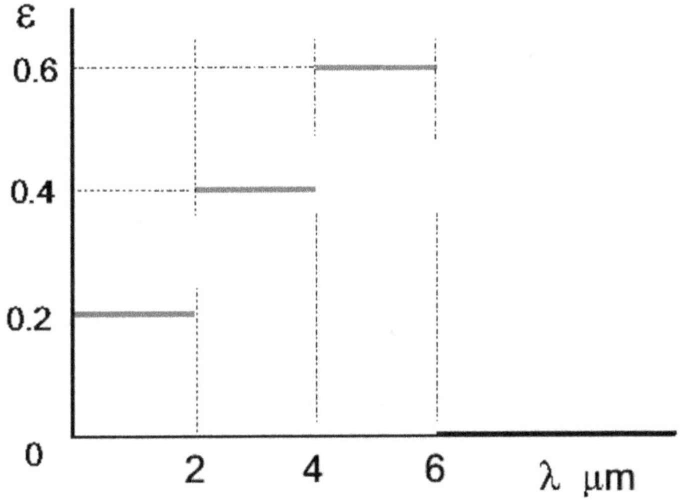

Figure 5.1. *Emissivity from the plate. For a color version of this figure, see www.iste.co.uk/ledoux/heat2.zip*

SOLUTION.–

1) In terms of radiation, we handle absolute temperatures:

$$T_W = 450 + 273 = 723\,K \qquad [5.75]$$

The emittance of a black body at temperature T_W would be:

$$M_T^0 = \sigma T_W^4 = 5.67.10^{-8} * 723^4 = 1.549.10^4\,W.m^{-2} \qquad [5.76]$$

Taking into account the variation in the emissivity with this wavelength, the emittance will be:

$$M_T = \sigma T_W^4 \left[0.2 * G_{0-2T} + 0.4 * G_{2T-4T} + 0.6 * G_{4T-6T}\right] \qquad [5.77]$$

We establish the table:

$\lambda\,\mu m$	$\lambda_T\ \mu m.K$	$G_{0-\lambda T}$	G_{2T-4T}	G_{4T-6T}
2	1446	0.01		
4	2892	0.2483	0.2383	
6	4338	0.5392		0.2909

And it becomes:

$$0.2*G_{0-2T}+0.4*G_{2T-4T}+0.6*G_{4T-6T}=0.2719 \quad [5.78]$$

$$M_T = 0.2719\sigma T_W^4 = 4212.5W.m^{-2} \quad [5.79]$$

The emittance is the energy emitted per surface unit. The flux emitted by a plate of $S = 10\,m^2$ will therefore be:

$$\Phi = S\,M_T = 4212.5*10 = 42125W \quad [5.80]$$

2) By defining the total emissivity:

$$\varepsilon = \frac{M_T}{\sigma T_W^4} \quad [5.81]$$

So:

$$\varepsilon = 0.2*G_{0-2T}+0.4*G_{2T-4}+0.6*G_{4T-6T}=0.2719 \quad [5.82]$$

NOTE.– It would have been possible to calculate ε directly using this formula.

3) We use Wien’s first law:

$$\lambda_m T_W = 2898\,\mu m.K \quad [5.83]$$

Hence:

$$\lambda_m = 4.008\,\mu m \qquad [5.84]$$

4) Yes, $\lambda_m = 4.008\,\mu m$ corresponds to maximum emissivity.

So $\lambda_m = 4.008\,\mu m$ is really the maximum emission.

5) The emissivity ε_λ is variable with the wavelength: the plate is not a gray body.

EXAMPLE 5.9.– Emissive plate: gray body modeling

A square side plate $a = 2m$ is kept at the temperature of $T = 75°\text{C}$.

Its emissivity ε_λ is equal to 0.8 in the wavelength band between $0.2\mu m$ and $20\mu m$; it is negligible on the rest of the spectrum.

1) What is the emittance M of this plate?

2) What radiative flux Φ_R escapes when it is at $T = 75°\text{C}$?

3) A busy engineer seeks to "model" this plate as a gray body. When it radiates at $T = 75°\text{C}$, what, therefore, will the emissivity ε_1 he will find be?

4) He re-does the calculation when the plate is at $T = 300°\text{C}$. What emissivity ε_2 will he find for this "gray body" modeling?

5) What do you think of the validity of this "gray body" modeling?

SOLUTION.–

1) We calculate the emittance M of the plate at $T = 75°\text{C}$?

$$T = 348\,K \qquad [5.85]$$

$$\lambda_1 T = 0.2 * 378 = 75.6\,\mu m.K \qquad [5.86]$$

$$G_{0-\lambda_1 T} \approx 0 \qquad [5.87]$$

$$\lambda_2 T = 20 * 378 = 7560 \mu m.K \quad [5.88]$$

$$G_{0-\lambda_2 T} = 0.8372 \quad [5.89]$$

$$M = \varepsilon \sigma T^4 \left(G_{0-\lambda_2 T} - G_{0-\lambda_1 T} \right) = 0.8 * 5.67.10^{-8} * (378)^4 * 0.8372$$
$$= 775.3W \quad [5.90]$$

2) Can the radiative flux Φ_R at 75°C be deduced directly from the emittance?

$$S = 2 * 2 = 4 m^2 \quad [5.91]$$

$$\Phi_R = S * M = 775.3 * 4 = 3101W \quad [5.92]$$

3) A busy engineer seeks to "model" this plate as a gray body. When it radiates at 75°C, what, therefore, will the emissivity ε_1 he will find be?

$$M = \varepsilon_1 \sigma T^4 \quad [5.93]$$

$$\varepsilon_1 = \frac{M}{\sigma T^4} = \frac{775.3}{5.67.10^{-8} T^4} = 0.67 \quad [5.94]$$

4) We calculate the emittance M of the plate at $T = 300°C$?

$$T = 573 K \quad [5.95]$$

$$\lambda_1 T = 0.2 * 573 = 114.6 \mu m.K \quad [5.96]$$

$$G_{0-\lambda_1 T} \approx 0 \quad [5.97]$$

$$\lambda_2 T = 20 * 573 = 11460 \mu m.K \quad [5.98]$$

$$G_{0-\lambda_2 T} = 0.938 \quad [5.99]$$

$$M = \varepsilon \sigma T^4 \left(G_{0-\lambda_2 T} - G_{0-\lambda_1 T} \right) = 0.8 * 5.67.10^{-8} * (573)^4 * 0.938 = 4587W \quad [5.100]$$

$$M = \varepsilon_2 \sigma T^4 \qquad [5.101]$$

$$\varepsilon_1 = \frac{M}{\sigma T^4} = \frac{4587}{5.67.10^{-8}\, T^4} = 0.75 \qquad [5.102]$$

5) The emissivity of the pseudo-gray body varies by 12% from one temperature to another. This modeling may only be suitable for a rough model.

EXAMPLE 5.10.– Surface at piecewise constant emissivity

A homogeneous surface temperature $T = 1327°C$ emits a diffuse radiation.

Its monochromatic emissivity ε_λ is given in the following table:

Range of λ in µm	**Monochromatic emissivity ε_λ**
$0-2\mu m$	$\varepsilon_\lambda = 0.4$
$2-5\mu m$	$\varepsilon_\lambda = 0.8$
Above $5\mu m$	$\varepsilon_\lambda = 0$

1) Determine the total hemispheric emissivity of this surface.

2) Determine the total emissive power of this surface.

3) For what wavelength is the radiation emission maximum?

SOLUTION.–

1) In such a problem, two strategies can be adopted:

Calculate the total emissive power of this surface (or total emittance) from the emissivity data, by combining the values of the function $G_{0-\lambda T}$ and Stefan's law. Then, from this, deduce the total emissivity from Stefan's law again. This amounts to interverting questions 1 and 2.

Or if you have more experience, calculate the total emissivity directly and, from this, deduce the total emissive power of this surface (or total emittance), which is asked here.

In terms of radiation, we know only absolute temperatures:

$$T = 1327 + 273 = 1600\,K$$

We must also first establish a table of energy fractions emitted in the wavelength bands given in the problem.

λ	λT	$G_{0-\lambda T}$	
$2\,\mu m$	$\lambda T = 3200\,\mu m.K$	$G_{0-\lambda T} = 0.3181$	
$5\,\mu m$	$\lambda T = 9000\,\mu m.K$	$G_{0-\lambda T} = 0.8889$	
$0-2\,\mu m$			0.3181
$2-5\,\mu m$			$0.8889 - 0.3181 = 0.5708$
$> 5\,\mu m$			$1 - 0.8889 = 0.111$

From this, we deduce the total emissivity ε

$$\varepsilon = (0.3181 * 0.4) + (0.5708 * 0.8) + (0.111 * 0) = 0.5839 \qquad [5.103]$$

2) Applying Stefan's law, we give the emissivity of the surface:

$$M = \varepsilon \sigma T^4 = 0.5839 * 5.67.10^{-8} * 1600^4 = 2.17.10^5\,W.m^{-2} \qquad [5.104]$$

3) The wavelength λ_{max} for which the emission is maximum is given by Wien's first law:

$$\lambda_{max} T = \; = 2898\,\mu m.K \qquad [5.105]$$

$$\lambda_{max} = \; = \frac{2898}{1600} = 1.81\,\mu m \qquad [5.106]$$

At this level, there is a problem. For this value of λ, the emissivity is ε = 0.4. From $\lambda = 2\,\mu m$, the emissivity becomes ε = 0.8. We can ask: how do we know where the maximum value is?

We know that, in any case, the black body luminance decreases with the wavelength above:

$$\lambda = 1.81\,\mu m. \qquad [5.107]$$

For two similar wavelengths, the black body radiation is proportional to λ^{-5}. The ratio of emittances from the black body at $\lambda = 1.81\,\mu m$ and $\lambda = 2\,\mu m$ will be

$$\frac{M^0_{1.81}}{M^0_2} \approx \left(\frac{2}{1.81}\right)^5 = 1.65 \qquad [5.108]$$

For the plate we are interested in, the emittances are given by that of the black body multiplied by the monochromatic emissivity. Therefore, the contribution at the emittance of the emission from the black body at $\lambda = 1.81$ μm is multiplied by 0.4 and that at $\lambda = 2$ μm is multiplied by 0.8. In other words, the ratio of contributions to emittance M_λ of wavelengths $\lambda = 1.81$ μm and $\lambda = 2$ μm will be:

$$\frac{M_{1.81}}{M_2} \approx \left(\frac{2}{1.81}\right)^5 \frac{0.4}{0.8} = 0.824 \qquad [5.109]$$

The emittance at $\lambda = 2$ *μm* will be higher at emittance $\lambda = 2$ *μm*. As the emittance decreases above $\lambda = 1.81$ *μm*, so from $\lambda = 2$ *μm*, the maximum emittance from the plate will be observed at

$$\lambda_{max} = 2\,\mu m \qquad [5.110]$$

The construction of the real emittance from the plate is illustrated by both Figures 5.2 and 5.3.

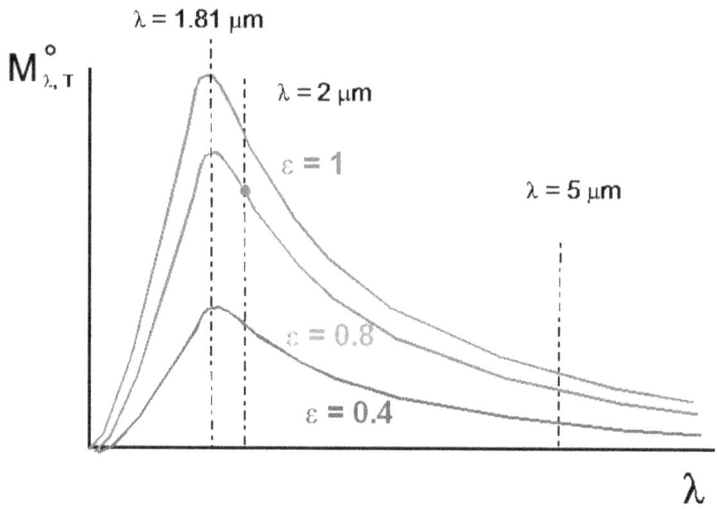

Figure 5.2. *Influence of the emissivity on the emittance from the plate. For a color version of this figure, see www.iste.co.uk/ledoux/heat2.zip*

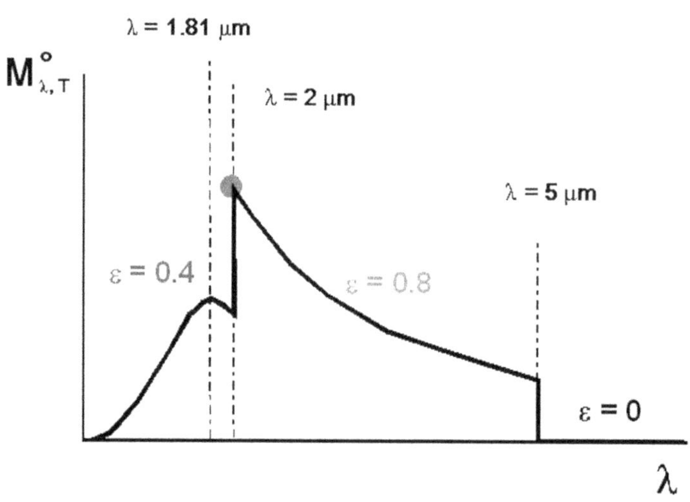

Figure 5.3. *Real emittance from the plate according to the wavelength. The maximum emission results in λ = 2 μm. For a color version of this figure, see www.iste.co.uk/ledoux/heat2.zip*

EXAMPLE 5.11.– A bright idea

An incandescent lamp is constructed by providing electrical power to a tungsten filament placed in a glass vacuum enclosure.

Tungsten melts at a temperature $T_f = 3422\,°\text{C}$. To preserve the filament, we project a limit of its operating temperature at $T_W = 3200\,°\text{C}$.

The filament has a length of $L = 30\,cm$. We seek to determine what its diameter d should be so that its temperature does not exceed the projected $T_W = 3200\,°\text{C}$.

A valid approximation will be to consider the tungsten as a gray body with emissivity $\varepsilon = 0.3$.

To be able to place a length $L = 30\,cm$ of wire in the bulb, the filament is rolled into a spiral; we will, however, consider that the thermal balance can be written for a straight wire.

The bulb is made of glass that is assumed to be perfectly transparent. The temperature of the room that the bulb illuminates is $T_a = 21\,°\text{C}$.

This filament is supplied with an electrical power $P = 75\,W$.

1) Write the thermal balance for the whole surface S of the filament.

2) From this, deduce the value of d so that the filament remains at the temperature $T_W = 3200°C$.

3) What is the maximum emission wavelength λ_{max} of the filament?

4) Using the energy supplied to the bulb

a) What fraction f_{vis}, expressed as a percentage of the power consumed by the bulb, will be used for lighting, that is, is in the visible spectrum between wavelengths $\lambda_1 = 0.35\,\mu m$ and $\lambda_2 = 0.75\,\mu m$?

b) Radiation known as "thermal" lies in the infrared spectrum between wavelengths $\lambda_2 = 0.75\,\mu m$ and $\lambda_3 = 100\,\mu m$.

What is the proportion $f_{thermal}$, expressed as a percentage of the electrical energy provided to the bulb, that serves only to heat the room?

SOLUTION.–

1) The emitting surface S has, for expression, neglecting the emission on the side of the filament, a very small area:

$$S = \pi\, d\, L \qquad [5.111]$$

The temperatures should be expressed in Kelvin:

$$T_W = 3200\,C = 3473\,K \qquad [5.112]$$

$$T_a = 21\,C = 294\,K \qquad [5.113]$$

The thermal balance at the surface of the filament should consider the electricity received per unit of time, P, heated by the Joule effect, the absorbed proportion of the radiation from the atmosphere, similar to a black body at temperature $T_a = 294K$, which is $S\,\varepsilon\,\sigma\,T_a^4$, both energies received and the radiated energy $S\,\varepsilon\,\sigma\,T_w^4$, the energy given at the exterior.

This balance, which is written for the surface of the filament, gives:

$$P + S\varepsilon\sigma T_a^4 = S\varepsilon\sigma T_w^4 \quad [5.114]$$

$$P = S\varepsilon\sigma\left(T_w^4 - T_a^4\right) \quad [5.115]$$

$$S = \frac{P}{\varepsilon\sigma\left(T_w^4 - T_a^4\right)} = \pi d L \quad [5.116]$$

2) All the parameters are known except *d*, which can be calculated as:

$$d = \frac{P}{\varepsilon\sigma\left(T_w^4 - T_a^4\right)\pi L} = \frac{75}{0.2 * 5.67.10^{-8} * \left(3473^4 - 294^4\right) * \pi * 0.3} \quad [5.117]$$

$$d = 3.216.10^{-5}\, m = 32.16\,\mu m \quad [5.118]$$

3) By applying Stefan's first law, we find:

$$\lambda_{\max} T_w = 2898\,\mu m.K \quad [5.119]$$

which is a maximum emission for the wavelength:

$$\lambda_{\max} = \frac{2898}{3473} = 0.83\,\mu m \quad [5.120]$$

which is just at the limit of visible infrared.

4) Let us make a table of the useful functions of $G_{0-\lambda T}$:

λ	λT_w	$G_{0-\lambda T_w}$
$\lambda = 0.35$	$\lambda T_w = 1215.5\,\mu m.K$	$G_{0-\lambda T_w} = 0.0024$
$\lambda = 0.35$	$\lambda T_w = 2604.7\,\mu m.K$	$G_{0-\lambda T_w} = 0.1831$
$\lambda = 100$	$\lambda T_w = 3.473.10^5\,\mu m.K$	$G_{0-\lambda T_w} = 1$

From this, we deduce the proportion of the energy emitted in the visible spectrum:

$$f_{vis} = 0.1831 - 0.0024 = 0.159 = 15.9\% \qquad [5.121]$$

From this, we deduce the proportion of the energy emitted in thermal infrared:

$$f_{thermal} = 1 - 0.1831 = 0.817 = 81.7\% \qquad [5.122]$$

We see that in such a bulb, nearly 82% of the energy is dissipated in thermal form.

5.3. Radiation and oven

EXAMPLE 5.12.– Black body oven

We equip a black body oven with a window cut in a material whose properties are the following:

Its transmission factor is $\tau = 0.8$ for wavelengths ranging from 0 to $3.5\,\mu m$, and $\tau = 0$ for all other wavelengths.

In emission and absorption, the window behaves like a gray body with emissivity $\varepsilon = 0.2$, so for all wavelengths from 0 to infinity.

The temperature of the oven is $T_f = 1500°\text{C}$.

In this problem, we are interested only in the light coming from the oven that is transmitted, absorbed or reflected by window. The emission from the window itself[1] *does not interest us here.*

1) What is the emittance M_T^0 from the oven?

2) What is the radiative flux density φ_{TR} transmitted by the window?

3) What is the flux density φ_{ABS} absorbed by the window?

1 For which the temperature, however, is not given.

4) From this, deduce the flux density φ_{REF} reflected by the window to the oven.

SOLUTION.–

1) The emittance M_T^0 from the oven is that of a black body given by Stefan's formula:

$$M_T^0 = \sigma T_F^4 \quad [5.123]$$

$$T_F = 1500C = 1773K \quad [5.124]$$

$$M_T^0 = 5.76.10^{-8} * 1773^4 = 5.6.10^5 \, Wm^{-2} \quad [5.125]$$

2) To determine the radiative flux density φ_{TR} transmitted by the window, we need to know the fraction of the black body radiation between 0 and 3.5 μm. This fraction will be given by the function:

$$\lambda T_F = 3.5 * 1773 = 6205 \quad [5.126]$$

$$G_{0-6205} = 0.7541 \quad [5.127]$$

With a transmission coefficient of $\tau = 0.8$, the flux density transmitted is:

$$\varphi_{TR} = \tau G_{0-6205} M_{TF}^0 \quad [5.128]$$

$$\varphi_{TR} = 0.8 * 0.7541 * 5.6.10^5 \quad [5.129]$$

$$\varphi_{TR} = 3.38.10^5 \, Wm^{-2} \quad [5.130]$$

3) The flux density φ_{ABS} absorbed by the window will be that of a gray body with an absorption coefficient equal to its emissivity:

$$\varphi_{ABS} = \varepsilon M_{TF}^0 \quad [5.131]$$

$$\varphi_{ABS} = 0.2 * 5.6.10^5 = 1.12.10^5 \quad [5.132]$$

4) The flux density φ_{REF} reflected by the window into the oven will be obtained through the difference between the incident flux, the part transmitted and the part absorbed:

$$\varphi_{REF} = M^0_{TF} - \varphi_{TR} - \varphi_{ABS} \quad [5.133]$$

$$\varphi_{REF} = 5.6.10^5 - 3.38.10^5 - 1.12.10^5 = 1.1.10^5\ Wm^{-2}$$

EXAMPLE 5.13.– Radiation provided by an oven

In an oven, a temperature $T_F = 2000°\text{C}$ is maintained.

This oven contains a perfectly transparent square window with sides of length $a = 2cm$.

1) What is the total emittance M_T from this window considered as a source of radiation? What is the flux Φ_R emitted by this window?

2) What is the wavelength λ_m at which the oven emits most?

3) We replace the window with a glass that transmits only some $\tau = 40\%$ of the light from the oven between, 0 and $1\mu m$, and transmits the entire remainder of the spectrum. Determine the emittance M'_T from this new window, as well as the flux Φ'_R that crosses it.

SOLUTION.–

1) The area of the window is:

$$S = a^2 = 4.10^{-4}\ m^2 \quad [5.134]$$

We can consider that the oven has a small aperture. We will therefore treat it as a black body. Stefan's law tells us

$$M_T = \sigma T_F^4 = 5.76.10^{-8} * (2000 + 273)^4 \quad [5.135]$$

$$M_T = 1.54.10^6\ W\ m^{-2} \quad [5.136]$$

The flux for

$$S = 4.10^{-4}\, m^2 \text{ will be} \qquad [5.137]$$

$$\Phi_R = S\, M_T = 4.10^{-4} * 1.54.10^{6} = 615\, W \qquad [5.138]$$

2) Wien's first law enables us to write:

$$\lambda_m\, T_F = 2898 \qquad [5.139]$$

Using the value 2,898 for the constant indicates that λ_m is expressed in μm and T_F in K.

Hence:

$$\lambda_m = \frac{2898}{T_F} = \frac{2898}{2273} = 1.275\, \mu m \qquad [5.140]$$

which lies in infrared.

3) Constant transmission window piecemeal

Between 0 and $1 \mu m$, the fraction of flux emitted by the oven is given by the function G_{0-1} whose value, given by the table, is $G_{0-1} = 0.1156$. Between $G_{0-1} = 0.1156$ and infinity, the fraction of radiation emitted by the oven over the rest of the spectrum will be:

$$1 - G_{0-1} = 1 - 0.1156 = 0.884 \qquad [5.141]$$

Adding the flux transmitted in each of the bands, we will have:

$$\Phi'_R = S\ \sigma T_F^4\, \tau\, G_{0-1} + S\ \sigma T_F^4 \left(1 - G_{0-1}\right) \qquad [5.142]$$

$$\Phi'_R = S\ \sigma T_F^4 \left[\tau\, G_{0-1} + \left(1 - G_{0-1}\right)\right]$$

$$= 4.10^{-4} * 5.67.10^{-8} * 2273^4 * \left(0.4 * 0.1156 + 0.884\right) \qquad [5.143]$$

$$\Phi'_R = 563\, W \qquad [5.144]$$

EXAMPLE 5.14.– Oven with transmission variable window

We equip a black body oven with a window cut from a material whose properties are the following:

Its transmission factor τ is 0.8 for wavelengths from 0 to 3.5 μm, and 0 for all the other wavelengths.

In emission and absorption, the window behaves as a gray body with emissivity $\varepsilon = 0.2$, so for all wavelengths from 0 to infinity.

The temperature of the oven is 1,500°C.

In this problem, we are interested only in the light coming from the oven that is transmitted, absorbed or reflected by the window. The emission from the window itself[2] *does not interest us here.*

1) What is the emittance M_T^0 from the oven?

$$M_T^0 = \sigma T_F^4 \quad [5.145]$$

$$T_F = 1500C = 1773K \quad [5.146]$$

$$M_T^0 = 5.67.10^{-8} * 1773^4 = 5.6.10^5 \, Wm^{-2} \quad [5.147]$$

2) What is the radiative flux density φ_{TR} transmitted by the window?

The fraction of black body radiation between 0 and 3.5μm will be given by the function:

$$\lambda T_F = 3.5 * 1773 = 6205 \quad [5.148]$$

$$G_{0-6205} = 0.7541 \quad [5.149]$$

With a transmission coefficient from 0.8, the flux density transmitted is:

$$\varphi_{TR} = \tau G_{0-6205} M_{TF}^0 \quad [5.150]$$

2 For which the temperature, however, is not given.

$$\varphi_{TR} = 0.8 * 0.7541 * 5.6.10^5 \qquad [5.151]$$

$$\varphi_{TR} = 3.38.10^5 \, Wm^{-2} \qquad [5.152]$$

3) Calculate the flux density φ_{ABS} absorbed by the window.

For a gray body, the emissivity is equal to the absorbance:

$$\alpha = \varepsilon \qquad [5.153]$$

$$\varphi_{ABS} = \varepsilon \, M_{TF}^0 \qquad [5.154]$$

$$\varphi_{ABS} = 0.2 * 5.6.10^5 = 1.12.10^5 \, W \, m^{-2} \qquad [5.155]$$

4) From this, deduce the flux density φ_{REF} reflected by the window into the oven

$$\varphi_{REF} = M_{TF}^0 - \varphi_{TR} - \varphi_{ABS} \qquad [5.156]$$

$$\varphi_{REF} = 5.6.10^5 - 3.38.10^5 - 1.12.10^5 = 1.1.10^5 \, Wm^{-2} \qquad [5.157]$$

EXAMPLE 5.15.– Black body oven 1

A cavity maintained at a uniform interior temperature of $T = 2000\,K$ has a small opening pierced into it.

1) Calculate the power of the radiation coming from this opening.

2) What is the wavelength λ_1 below which 10% of the radiation from the cavity is concentrated?

3) What is the wavelength λ_2 below which 90% of the radiation from the cavity is concentrated?

4) What is the wavelength λ_{max} associated with maximum power?

SOLUTION.–

1) The oven is a black body.

If s is the surface of the opening, the power is the energy leaving per unit of time.

$$M_0 = \sigma T^4 \quad [5.158]$$

$$P = M_0 \, s = s \sigma T^4 \quad [5.159]$$

$$M_0 = \sigma T^4 = 5.67.10^{-8} * 2000^4 = 9.072 10^5 \, W.m^{-2} \quad [5.160]$$

2) In the tables, we read

$$G_{0-\lambda T} = 0.1009 \text{ for } \lambda T = 2200 \, \mu m.K \quad [5.161]$$

which makes the wavelength sought:

$$\lambda_1 = \frac{2200}{2000} = 1.1 \, \mu m \quad [5.162]$$

3) In the tables, we read

$$G_{0-\lambda T} = 0.9 \text{ for } \lambda T = 9400 \, \mu m.K \quad [5.163]$$

which makes the wavelength sought:

$$\lambda_2 = \frac{9400}{2000} = 4.7 \, \mu m \quad [5.164]$$

4) This wavelength λ_{max} is given by Wien's first law:

$$\lambda_{max} T = 2898.10^{-6} \, m.K = 2898 \, \mu m.K \quad [5.165]$$

$$\lambda_{max} = \frac{2898.10^{-6} \, \mu m.K}{2000} = 1.449 \, \mu m \quad [5.166]$$

EXAMPLE 5.16.– Black body oven 2

In a “black body” oven, there is a temperature of 1,327°C.

This oven is equipped with a square window with sides of length $a = 5\,cm$.

This window is covered by two filters:

A “high-pass” filter that allows only wavelengths higher than $\lambda_1 = 1\,\mu m$ to pass and a “low-pass” filter that allows only wavelengths lower than $\lambda_2 = 5\,\mu m$ to pass.

1) What is the emittance M from the radiation that crosses the window?

2) What is the flux Φ that crosses this window?

SOLUTION.–

1) The emittance is the flux per surface unit that crosses the window. This is the fraction of the emission from a black body between $\lambda_1 = 1\,\mu m$ and $\lambda_2 = 5\,\mu m$.

It has

$$T = 1327 + 273 = 1600\,K \qquad [5.167]$$

$\lambda_1 T = 1600\,\mu m.K$	$G_{0-\lambda_1 T} = 0.0197$	$G_{\lambda_1 T-\lambda 2_1 T} = 0.836$
$\lambda_2 T = 8000\,\mu m.K$	$G_{0-\lambda_2 T} = 0.8562$	

It has

$$M = G_{\lambda_1 T-\lambda 2_1 T}\,\sigma T^4 \qquad [5.168]$$

$$M = 0.836 * 5.67.10^{-8} * 1600^4 \qquad [5.169]$$

$$M = 3.106.10^5\,W.m^{-2} \qquad [5.170]$$

$$S = 5.10^{-2} * 5.10^{-2} = 25.10^{-4}\,m^2 \qquad [5.171]$$

2) The flux is the energy that crosses the whole surface of the window.

$$\Phi = S\,M_T = S\,G_{\lambda_1 T-\lambda 2_1 T}\,\sigma T^4 \qquad [5.172]$$

$$\Phi = 25.10^{-4} * 3.106.10^5 \qquad [5.173]$$

$$\Phi = 776.6W \qquad [5.174]$$

EXAMPLE 5.17.– Radiation transmission by glass

A glass plate with a surface $S = 100cm^2$ is used to observe the radiation coming from an oven. The transmission factor τ from the glass is null except in the band $\lambda_1 = 0.2\mu m$ at $\lambda_2 = 3.5\mu m$ where it amounts to $\tau = 0.8$. The emission factor is taken to be equal to $\varepsilon_1 = 0.2$ to $\lambda_2 = 3.5\mu m$ and $\varepsilon_2 = 0.9$ beyond that.

By admitting that the oven is a black body at $T = 1800°C$, calculate the total energy absorbed by the glass and the energy transmitted.

SOLUTION.– First, calculate the fractions of the energy transmitted by the oven black body in different bands of absorption and transmission, which is useful for this problem. We will use the functions $G_{0-\lambda T}$ as usual.

The temperature of the oven is

$$T = 1800°C = 2073\,K \qquad [5.175]$$

Construct this table:

$\lambda_1 = 2\mu m$	$\lambda_1 T = 414.6\mu m.K$	$G_{0-\lambda_1 T} \approx 0$
$\lambda_2 = 3.5\mu m$	$\lambda_2 T = 7256\mu m.K$	$G_{0-\lambda_2 T} = 0.822$
Energy emitted beyond $\lambda_2 = 3.5\mu m$		$1 - 0.822 = 0.178$

The fraction of the energy transmitted f_τ will be:

$$f_\tau = \tau\left(G_{0-\lambda_2 T} - G_{0-\lambda_1 T}\right) = 0.8 * (0.822 - 0) \qquad [5.176]$$

$$f_\tau = 0.658 \qquad [5.177]$$

The fraction of energy absorbed f_{abs} will be the sum of the energy contained in the spectrum $0 - 3.5\,\mu m$, absorbed at 20%, and of the energy contained in the remainder of the spectrum, absorbed at 90%

$$f_{abs} = (0.2 * 0.822) + (0.9 * 0.178) = 0.325 \qquad [5.144]$$

The fraction reflected f_R by the window into the oven will be:

$$f_R = 1 - (f_\tau + f_{abs}) = 1 - (0.658 + 0.325) \qquad [5.144]$$

$$f_R = 0.017 \qquad [5.144]$$

5.4. Radiation and metrology

EXAMPLE 5.18.– Measuring temperature. Color temperature

We wish to determine the temperature of a block of radiating metal by measuring the ratio of its luminances L_1 and L_2, respectively, for two wavelengths λ_1 and λ_2. We assume that in the range of wavelengths measured, Wien's law is a valid approximation of the expression of the black body luminance.

We make measurements in infrared; we choose $\lambda_1 = 1.6\,\mu m$ and λ_2=1 μm.

We measure a luminance ratio $\dfrac{L_2}{L_1} = 150$.

1) We make the hypothesis that the metal is a black body.

The temperature thus determined is called a **color temperature**.

What is the color temperature T_C of the block?

2) We make the hypothesis that the metal is a gray body.

a) Show that we do not need to hypothesize on the emissivity ε of this gray body.

b) What new temperature is thus determined?

3) We finally know the emissivities of the metal for two wavelengths λ_2 and λ_1, which are, respectively, ε_2 and ε_1.

a) What is the real temperature T_R of the metal?

b) What relative error represents assimilation of the color temperature to the true temperature of the metal?

4) Does the approximation of Planck's luminance law with Wien's law seem legitimate to you?

SOLUTION.–

1) Using Wien's law:

$$L^0(\lambda,T) \approx C_1\,\lambda^{-5}\exp\left(\frac{-C_2}{\lambda T}\right) \qquad [5.178]$$

Making the hypothesis of a black body, emissivities at λ_2 and λ_1 are equal to 1.

The ratio of the luminances at λ_2 and λ_1 is therefore expressed as:

$$\frac{L_2}{L_1} = \left(\frac{\lambda_1}{\lambda_2}\right)^5 \exp\left[\frac{-C_2}{T_C}\left(\frac{1}{\lambda_2}-\frac{1}{\lambda_1}\right)\right] \qquad [5.179]$$

We will extract T_C

$$T_C = \frac{-C_2\left(\dfrac{1}{\lambda_2} - \dfrac{1}{\lambda_1}\right)}{Ln\left[\dfrac{L_2}{L_1}\left(\dfrac{\lambda_1}{\lambda_2}\right)^{-5}\right]} \qquad [5.180]$$

which is:

$$T_C = \frac{-1.4388.10^{-2}\left(\dfrac{10^6}{\lambda_2} - \dfrac{10^6}{\lambda_1}\right)}{Ln\left[0.731*(1.6)^{-5}\right]} = \frac{-5395.5}{-2.663} \qquad [5.181]$$

$$T_C = 2026\,K = 1753°\text{C} \qquad [5.182]$$

2) Making the hypothesis of a gray body, emissivities at λ_2 and λ_1 are equal to a single value ε. The luminance becomes:

$$L^0\,(\lambda,T) \approx \varepsilon C_1\,\lambda^{-5}\exp\left(\frac{-C_2}{\lambda T}\right) \qquad [5.183]$$

The ratio of the luminances at λ_2 and λ_1 is therefore expressed as:

$$\frac{L_2}{L_1} = \left(\frac{\lambda_1}{\lambda_2}\right)^5 \frac{\varepsilon}{\varepsilon}\exp\left[\frac{-C_2}{T}\left(\frac{1}{\lambda_2} - \frac{1}{\lambda_1}\right)\right] \qquad [5.184]$$

a) The emissivity only appears in a ratio equal to 1. There is therefore no point in giving it a value.

b) The temperature found will remain the same. We therefore find the color temperature:

$$T_C = 2026\,K = 1753°\text{C} \qquad [5.185]$$

3) We finally know the emissivities of the casting for both wavelengths λ_2 and λ_1, which are, respectively,

$$\varepsilon_2 = 0.4 \text{ and } \varepsilon_1 = 0.35 \qquad [5.186]$$

a) The luminances become:

$$L_1\,(\lambda_1,T) \approx \varepsilon_1\, C_1\, \lambda_1^{-5} \exp\left(\frac{-C_2}{\lambda_1\, T_R}\right) \qquad [5.187]$$

$$L_2\,(\lambda_2,T) \approx \varepsilon_2\, C_1\, \lambda_2^{-5} \exp\left(\frac{-C_2}{\lambda_2\, T_R}\right) \qquad [5.188]$$

And their ratio:

$$\frac{L_2}{L_1} = \left(\frac{\lambda_1}{\lambda_2}\right)^5 \frac{\varepsilon_2}{\varepsilon_1} \exp\left[\frac{-C_2}{T_R}\left(\frac{1}{\lambda_2} - \frac{1}{\lambda_1}\right)\right] \qquad [5.189]$$

and T_R is calculated from:

$$T_R = \frac{-C_2\left(\dfrac{1}{\lambda_2} - \dfrac{1}{\lambda_1}\right)}{Ln\left[\dfrac{L_2}{L_1}\dfrac{\varepsilon_2}{\varepsilon_1}\left(\dfrac{\lambda_1}{\lambda_2}\right)^{-5}\right]} \qquad [5.190]$$

$$T_R = \frac{-1.4388.10^{-2}\left(\dfrac{10^6}{\lambda_2} - \dfrac{10^6}{\lambda_1}\right)}{Ln\left[0.731*1.43*(1.6)^{-5}\right]} = \frac{-5395.5}{-3.804} \qquad [5.191]$$

$$T_R = 1750\,K = 1477°\text{C} \qquad [5.192]$$

b) The relative error on the temperature measurement is:

$$\frac{\Delta T_R}{T_R} = \frac{2026 - 1750}{1750} = 15.8\% \qquad [5.193]$$

4) Wien's approximation identifies the term in $\dfrac{1}{\exp\left(\dfrac{C_2}{\lambda_1 T}\right)-1}$ of Planck's law at the exponential $\exp\left(\dfrac{-C_2}{\lambda T}\right)$, which means considering $\exp\left(\dfrac{-C_2}{\lambda T}\right)$ as larger than 1

$$L^0(\lambda,T) \approx C_1 \lambda^{-5} \exp\left(\frac{-C_2}{\lambda T}\right) \qquad [5.194]$$

where

$$C_1 = 2hc^2 = 1.19.10^{-16}\, W.m^2 \qquad [5.195]$$

and

$$C_2 = \frac{hc}{k} = 1.4388.10^{-2}\, m.K = 14388\, \mu m.K \qquad [5.196]$$

we have

$$\lambda_1 T = 1.6.10^{-6} * 2000 = 3.2.10^{-3} \qquad [5.197]$$

and

$$\lambda_1 T = 10^{-6} * 2000 = 2.10^{-3} \qquad [5.198]$$

which can be considered small. We seem to be returning to the condition needed to validate Wien's law:

$$\lambda T \approx \varepsilon. \qquad [5.199]$$

We can be more confident by comparing the exponential $\exp\left(\dfrac{-C_2}{\lambda T}\right)$ to 1:

We thus calculate:

$$\exp\left(\frac{-C_2}{\lambda_1 T}\right)=\exp\left(\frac{-1.4388.10^{-2}}{1.6.10^{-6}*1750}\right)=5.87.10^{-3} \qquad [5.200]$$

From this, we draw:

$$\exp\left(\frac{C_2}{\lambda_1 T}\right)=170.47 \qquad [5.201]$$

which is greater than 1.

Above all, the **approximate** term can be determined **by** $\exp\left(\frac{-C_2}{\lambda_1 T}\right)$ in Planck's law:

$$\frac{1}{\exp\left(\frac{C_2}{\lambda_1 T}\right)-1}=5.90.10^{-3}\approx\exp\left(\frac{-C_2}{\lambda_1 T}\right) \qquad [5.202]$$

The degree of approximation of Wien's law is here in the order of 5 for 1,000.

EXAMPLE 5.19.– Measuring the temperature in an oven

Prologue

The thesis supervisor of the author of this problem often said: *"There are some engineers who need to understand that you cannot measure the temperature in an oven with a chair bar"*.

This pictorial reflection inspired what follows.

Memorandum on basic ideas

a) A thermocouple results from soldering two threads of different metals. A tension proportional to the temperature of the soldering is then delivered. This thermocouple manufactured in a laboratory takes the form of a sphere with the diameter d. The indication given by a thermocouple is therefore the temperature of this sphere.

b) Thermal exchanges between one sphere and the ambient air

We know that a sphere plunged into a gaseous atmosphere can exchange heat with the atmosphere via conduction and convection. The convection coefficient (including conductive phenomena) is then deduced from the expression, written by Wittaker:

$$N_u = 2 + \left[0.4\, R_D^{1/2} + 0.06\, R_D^{2/3} \right] P_r^{0.4}, \qquad [5.203]$$

where $R_D = \dfrac{U d}{\nu}$ is the Reynolds number of the flow and $P_r = \dfrac{\nu}{a}$ is the Prandtl gas number. The term equal to 2 results from the contribution from conduction. In a calm atmosphere, only the term $N_u = 2$ remains.

N_u is a number called the *Nusselt* number, defined by the convection coefficient h, by the thermal conductibility of the oven gasses λ, and the diameter d of the sphere:

$$N_u = \frac{hd}{\lambda} \qquad [5.204]$$

Following processes that will not be detailed here, the temperature of the gasses in the oven is constant and equal to $T_G = 900°\mathrm{C}$.

It is this temperature T_G that we wish to measure by plunging the thermocouple in the atmosphere of the oven.

The walls of the oven act as a gray body at temperature T_F. The gaseous atmosphere in the oven is perfectly calm.

The gasses in the oven are considered to be **perfectly transparent. They are not the subject of any radiation transfer. Only the thermocouple and the walls of the oven are the subject of radiation transfers.**

However, we should consider conductive transfers from the thermocouple with the atmosphere.

The thermocouple is placed at a temperature T_w, which will therefore be the result of the temperature measured performed. T_w, we will see, is lower than T_G.

The thermocouple is considered to be a gray body with emissivity $\varepsilon = 0.4$. We give the thermal conductibility of gasses in the oven $\lambda = 0.16\ W.m^{-1}.K^{-1}$. The real temperature of the oven walls is $T_F = 1400$°C.

1) Express the value of h depending on the diameter d of the thermocouple.

2) Write the thermal conduction and radiation balance on a surface unit of the thermocouple. From this, deduce the equation that the temperature T_w meets.

3) Results of a measurement

a) Calculate the temperature measured T_{w1} with a thermocouple with diameter $d_1 = 100\,\mu m$. What is the relative error?

b) Calculate the temperature measured T_{w2} with a thermocouple with diameter $d_2 = 1\,mm$. What is the relative error?

c) What do you think of the comparison of the chair bar in metrology?

4) Surface treatment

We again take the measurement of a thermocouple of diameter $d_2 = 1\,mm$, but we process the surface so that its emissivity becomes $\varepsilon = 0.9$. What becomes the value measured and the relative error?

SOLUTION.–

1) The "convection" coefficient is given by:

$$N_u = \frac{hd}{\lambda} = 2 \qquad [5.205]$$

which is:

$$h=\frac{\lambda N_u}{d}=\frac{2\lambda}{d}=\frac{0.32}{d}$$ [5.206]

2) The temperatures should be expressed in Kelvin:

$$T_F=1400°\text{C}=1673\,K$$ [5.207]

$$T_G=900°\text{C}=1173\,K$$ [5.208]

The balance per surface unit is built from:

– the radiation emitted by the black body oven and absorbed by the thermocouple, which is $\varepsilon\sigma T_f^4$;

– the radiation emitted by the thermocouple, which is $\varepsilon\sigma T_w^4$;

– the conduction between the gas and the thermocouple, which is $h(T_W-T_G)$.

NOTE.– We do not know at the outset whether T_W is greater or less than T_F.

The balance is then written in each case as:

– if $T_G>T_W$

$$\varepsilon\sigma T_F^4+h(T_G-T_W)=\varepsilon\sigma T_w^4$$ [5.209]

– if $T_G<T_W$ [5.210]

$$\varepsilon\sigma T_F^4=\varepsilon\sigma T_w^4+h(T_W-T_G)$$ [5.211]

In each case, the formal expression of the equation is the same. So, the solution tells us in which direction the conductive transfers go:

$$\varepsilon\sigma\left(T_F^4-T_w^4\right)+h(T_G-T_W)=0$$ [5.212]

We see that $T_F > T_W$, so $T_F - T_W > 0$; there result $T_G - T_W < 0$ and $T_G < T_W$. [5.213]

The temperature measured will be higher than the real temperature of the gasses.

NOTE.– In this case, it is not pertinent to use the approximation:

$$T_w^4 - T_F^4 \approx 4T_F^3\left(T_W - T_F\right) \quad [5.214]$$

The equation is linearized by:

$$\varepsilon\sigma 4T_F^3\left(T_F - T_W\right) + h\left(T_G - T_W\right) = 0 \quad [5.215]$$

$$\varepsilon\sigma 4T_F^3\left(T_F - T_W\right) + \frac{2\lambda}{d}\left(T_G - T_W\right) = 0 \quad [5.216]$$

for $d = 50\,\mu m$, the equation is written as:

$$425*\left(1673 - T_W\right) + \frac{0.32}{50.10^{-6}}\left(1173 - T_W\right) = 0 \quad [5.217]$$

$$\left(425 + 6400\right)T_W - \left(2098 + 7.5.10^6\right) \quad [5.218]$$

$$T_W = 1099 < 1173\ !!!!! \quad [5.219]$$

We note that $T_W - T_F = 900 - 1673 = 773$ is a long way from being smaller than $T_F = 1673$!

$$\varepsilon\sigma\left(T_F^4 - T_w^4\right) + h\left(T_G - T_W\right) = 0 \quad [5.220]$$

$$\varepsilon\sigma T_w^4 + \frac{2\lambda}{d}T_W - \left(\varepsilon\sigma T_F^4 + \frac{2\lambda T_G}{d}\right) = 0 \quad [5.221]$$

$$2.27.10^{-8}\,T_w^4 + \frac{0.32}{d}T_W - \ 1.777.10^5 + \frac{375.4}{d} \ = 0 \quad [5.222]$$

for $d = 50\,\mu m$, the equation is written as:

$$2.27.10^{-8}\,T_w^4 + 6400\,T_W - \left(1.777.10^5 + 7.508.10^6\right) = 0 \qquad [5.223]$$

$$2.27.10^{-8}\,T_w^4 + 6400\,T_W - \left(7.686.10^6\right) = 0 \qquad [5.224]$$

$$T_W = 1193\,K = 920°\text{C} \qquad [5.225]$$

3) The relative error on the temperature of the gasses is:

$$\frac{920 - 900}{900} = 2.2\% \qquad [5.226]$$

a) for $d = 100\,\mu m,$ the equation is written as:

$$2.27.10^{-8}\,T_w^4 + 3200\,T_W - \left(1.777.10^5 + 3.75.10^6\right) = 0 \qquad [5.227]$$

$$2.27.10^{-8}\,T_w^4 + 3200\,T_W - \left(3.928.10^6\right) = 0 \qquad [5.228]$$

$$T_W = 1212.18\,K = 939.2°\text{C} \qquad [5.229]$$

The relative error on the temperature of the gasses is:

$$\frac{1212.8 - 900}{900} = 4.35\% \qquad [5.230]$$

b) for $d = 1mm,$ the equation is written as:

$$2.27.10^{-8}\,T_w^4 + 320\,T_W - \left(1.777.10^5 + 3.754.10^5\right) = 0 \qquad [5.231]$$

$$2.27.10^{-8}\,T_w^4 + 320\,T_W - \left(5.531.10^5\right) = 0 \qquad [5.232]$$

$$T_W = 1430\,K = 1157°\text{C} \qquad [5.233]$$

The relative error on the temperature of the gasses is:

$$\frac{1157-900}{900}=28.5\% \qquad [5.234]$$

c) This places the "metrological chair bar" at $d=1mm$.

4) For $d=1mm$, but by taking the emissivity of the thermocouple to $\varepsilon=0.9$, the equation is written as:

$$5.1.10^{-8}T_w^4+320T_W-\left(4.10^5+3.754.10^5\right)=0 \qquad [5.235]$$

$$5.1.10^{-8}T_w^4+320T_W-\left(7.754.10^5\right)=0 \qquad [5.236]$$

$$T_W=1536\,K=1263°\text{C} \qquad [5.237]$$

The relative error on the temperature of the gasses is:

$$\frac{1263-900}{900}=40\% \qquad [5.238]$$

EXAMPLE 5.20.– Influence of radiation on temperature measurements

A thermo sensor of diameter d_t with an emissivity $\varepsilon_1=0.8$ is used to measure the temperature of a transparent gas flowing into an enclosure whose walls are at $T_a=300°\text{C}$. The real temperature of the gas is $T_G=600°\text{C}$.

1) Calculate the temperature T_{Meas1} indicated by the sensor, given that the convective exchange coefficient is $h=122W.m^{-2}K^{-1}$.

2) What does the measured temperature, which is $T_{Meas1'}$, become if the sensor is covered in a "coating" with emissivity $\varepsilon_2=0.1$?

SOLUTION.–

1) As in any radiation problem, let us first express the temperatures in Kelvin:

$$T_a = 300°\text{C} = 573\,K \quad [5.239]$$

$$T_G = 600°\text{C} = 873\,K \quad [5.240]$$

The surface of the sensor is in thermal equilibrium at temperature T_{Meas1} under the influence of different transfers, which is per surface unit of the sensor:

– a convective flux density

$$\varphi_{conv} = h\left(T_{Meas1} - T_G\right) \quad [5.241]$$

– an absorbed radiation φ_{RAbs} coming from the walls of the enclosure that behave as a black body. Considering the emissivity $\varepsilon_1 = 0.8$ equal to the absorption coefficient for this gray body

$$\varphi_{RAbs} = \varepsilon_1\,\sigma T_a^4 \quad [5.242]$$

– a radiation emitted by the sensor at T_{Meas1}. Considering the emissivity $\varepsilon_1 = 0.8$ for this gray body

$$\varphi_{REmitt} = \varepsilon_1\,\sigma T_W^4 \quad [5.243]$$

At this stage, we do not know if the equilibrium temperature where the sensor T_{Meas1} is placed is higher than T_{Meas1} or not.

If $T_{Meas1} > T_G$, the convective flux leaves the sensor and the balance is written as:

$$\varepsilon_1\,\sigma T_a^4 = h\left(T_{Meas1} - T_G\right) + \varepsilon_1\,\sigma T_{meas1}^4 \quad [5.244]$$

If $T_{Meas1} < T_G$, the convective flux leaves the sensor and the balance is written as:

$$\varepsilon_1 \sigma T_a^4 + h(T_G - T_{Meas1}) = \varepsilon_1 \sigma T_{Meas1}^4 \quad [5.245]$$

We see that in the two hypotheses, the equation to be solved is the same (the sign of the convective term adjusting the balance):

$$h(T_{Meas1} - T_G) + \varepsilon_1 \sigma \left(T_{Meas1}^4 - T_a^4 \right) = 0 \quad [5.246]$$

which is:

$$4.538.10^{-8} T_{Meas1}^4 + 122 T_{Meas1} - 1.065.10^5 - 4890$$

$$= 4.538.10^{-8} T_{Meas1}^4 + 122 T_{Meas1} - 1.1139.10^5 = 0 \quad [5.247]$$

Using a solver, common on scientific calculators, we give $T_{Meas1} = 777.27$.

Compared to the temperature of the gasses, which is $T_G = 873 K$, the measurement error is

$$\Delta T_{Meas1} = 873 - 777.27 = 95.73 \, K \quad [5.248]$$

which is

$$\frac{\Delta T_{Meas1}}{T_G} = 11\% \quad [5.249]$$

2) When the sensor is covered in a coating, the new value of the emissivity leads to a balance of the equation:

$$h(T_{Meas1} - T_G) + \varepsilon_2 \sigma \left(T_{Meas1}^4 - T_a^4 \right) = 0 \quad [5.250]$$

which is

$$6.57.10^{-9} T_{Meas1}^4 + 122 T_{Meas1} - 1.065.10^5 - 6112$$

$$= 6.57.10^{-9} T_{Meas1}^4 + 122 T_{Meas1} - 1.126.10^5 = 0 \quad [5.251]$$

Using a solver common on scientific calculators gives us T_{Meas2}=889.27.

Compared to the temperature of the gasses, which is T_G = $873K$, the measurement error is = $\Delta T_{Meas2} = 873 - 889.27 = -16.27\ K$, which is

$$\frac{\Delta T_{Meas2}}{T_G} = 1.9\% \quad [5.252]$$

5.4.1. *Measuring a thermal conductibility*

We seek to measure the conductibility λ of a material by placing a cylinder of length $L = 10\,cm$ and diameter $D = 2\,cm$ between two copper plates P_1 and P_2 of the same diameter $D = 2\,cm$ as the sample to be measured.

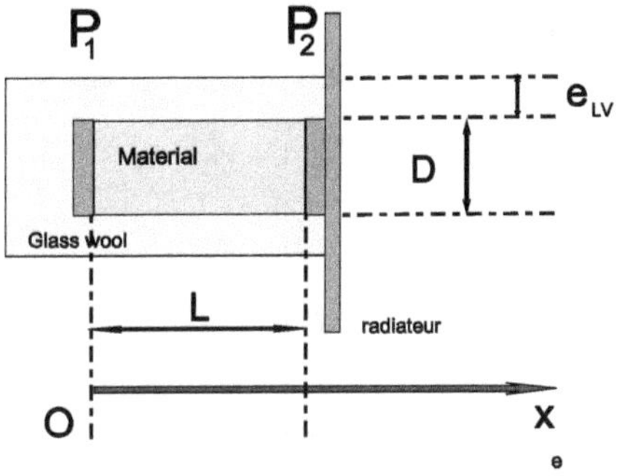

Figure 5.4. *Device for measuring thermal conductibility. First time: no error source is identified. For a color version of this figure, see www.iste.co.uk/ledoux/heat2.zip*

The plate P_1 is equipped on its face in contact with the sample from a thermocouple deposited by evaporation (thus not forming any excess thickness). We consider moreover that the plates are at a homogeneous temperature, respectively T_1 and T_2. The plate P_2 is soldered to a radiator formed of a massive plate. This plate is flat, circular, of diameter $D_R = 50\,cm$ and forms with plate P_2 a block isothermic to the temperature T_2, which is $T_a = 19°C$, the temperature of the laboratory.

The system is surrounded by a substantial mass of thermically isolating glass wool. In P_1, an electrical resistance delivers a power P.

For a power $P = 25W$, we measure T_1 =233°C as well as T_a =19°C. These are the measurement values that will be used for the following.

1) In an initial case, we assume that:

– the glass wool is perfectly isolating;

– the contact between the sample and the plates P_1 and P_2 is perfect;

– the radiator is efficient enough for the temperature T_2 to be equal to the ambient temperature T_a.

a) What is the heat flux φ and the heat flux density φ_0 delivered by the plate P_1?

b) What is the result of the conductibility measurement of the material, λ_{1mes}, using these hypotheses?

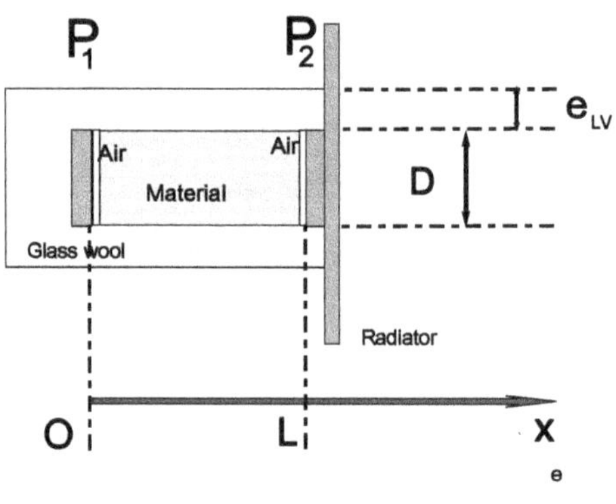

Figure 5.5. *Device for measuring thermal conductibility a second time. An initial source of error: air gaps. For a color version of this figure, see www.iste.co.uk/ledoux/heat2.zip*

2) A second time, we assume that

– the glass wool is perfectly isolating;

– the radiator is efficient enough for the temperature T_2 to be equal to the ambient temperature T_a;

– on the contrary, contact between the sample and the plates P_1 and P_2 is not perfect: there is an air gap of thickness $e = 10\,\mu m$ between each plate and the ends of the material.

We recall the thermal conductibility in air: $\lambda_{air} = 0.024\,W.m^{-1}.K^{-1}$.

a) What is the thermal resistance R_{thair} of an air gap? What difference in temperature appears between each plate and the material?

b) Considering the presence of air gaps, reinterpret the measurement using these hypotheses.

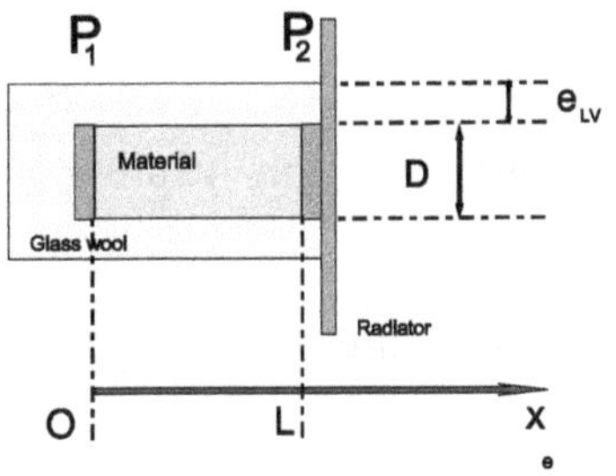

Figure 5.6. *Device for measuring thermal conductibility. Third time: a second source of error, the real temperature of the radiator. For a color version of this figure, see www.iste.co.uk/ledoux/heat2.zip*

3) A third time, we assume that

– the glass wool is perfectly isolating;

– contact between the sample and the plates P_1 and P_2 is perfect;

– on the contrary, we consider the fact that the radiator, to evacuate the power, should be at a temperature T_{rad} significantly higher than T_a. This radiator is considered a gray body with emissivity $\varepsilon = 0.9$. Moreover, we consider that the heat also leaves by reason of a convection coefficient $h = 5\,W.m^{-2}.K^{-1}$.

a) Calculate T_{rad}.

b) We define an "apparent" thermal resistance R_{app} by:

$$\varphi = \frac{T_{Rad} - T_a}{R_{app}} \qquad [5.253]$$

Here, what is the value found for R_{app}?

c) We try to justify the definition of this apparent thermal resistance by using the approximation:

$$T_w^4 - T_a^4 \approx 4T_a^3 \left(T_W - T_a\right)$$

What then is the expression found for R_{app}?

Compare the value found with this expression to that found in b). Comment.

d) From this, deduce the measurement for conductibility of the material, λ_{3mes}, using these hypotheses.

We recall the value of Stefan's constant $\sigma = 5.67.10^{-8}\ SI$.

4) Finally, we use the measurement without making any more simplifying hypotheses, which leads to considering both sources of error raised above:

– contact between the sample and the plates P_1 and P_2 is not perfect, there is an air gap of thickness $e = 5\,\mu m$ between each plate and the ends of the material;

– we consider the fact that the radiator, to evacuate the power, should be at a temperature T_2, significantly higher than T_a.

What therefore is the value found for λ? Taking the value as an "exact" measurement reference λ_{real}, what is the relative error committed on the measurement in the four previous cases?

5) At the start of the experiment, we place tension on the resistance present in P_1.

We estimate that the experiment is carried out correctly when the temperature measured on P_1 is equal to 99% of the final value T_1. The density of the material is $\rho = 7833\ kg.m^{-3}$, and its calorific capacity is $c = 465\ J.kg^{-1}$. After how long can we reasonably measure temperature?

SOLUTION.–

1) The surface of the sample and the plates is

$$S = \pi \frac{D^2}{4} = 3.14.10^{-4}\ m^2 \qquad [5.254]$$

a) The flux delivered by the plate is $\varphi = P = 25W$.

The flux density delivered by the plate is therefore:

$$\varphi_0 = \frac{P}{S} = 79\ 618\ W.m^{-2} \qquad [5.255]$$

b) Applying Fourier's law without caution, with $T_2 \equiv T_a = 19°C$

$$\varphi_0 = \lambda_{meas1} \frac{T_1 - T_a}{L} = 79618W.m^{-2} \qquad [5.256]$$

$$79618 = \lambda_{meas1} \frac{233 - 19}{0.1} \qquad [5.257]$$

So, for λ_{meas1}, we will measure the value:

$$\lambda_{meas1} = 37.2W.m^{-1}.K^{-1} \qquad [5.258]$$

2) Presence of air gaps

a) Thermal resistance per m² of the air gap is:

$$R_{thair} = \frac{5.10^{-6}}{0.024} = 2.08.10^{-4} \quad [5.259]$$

Thermal resistance from the bar is:

$$R_{thbar} = \frac{0.1}{\lambda} \quad [5.260]$$

The jump in temperature caused by an air gap will be:

considering the flux density, each of these two air gaps causes a difference in parasite temperature of

$$\Delta T_{air} = R_{thair}\,\varphi_0 = 16.6°\,C\,. \quad [5.261]$$

b) Applying Fourier's law, by recalculating the value of $T_1 - T_2$ with $T_2 \equiv T_a = 19°\,C$.

The real difference in temperature between two faces of the bar, in this model, is:

$$T_1 - T_2 = 223 - 19 - 2*16.6 = 170.8°C \quad [5.262]$$

$$\varphi_0 = \lambda_{meas1}\frac{T_1 - T_2}{L} = 79\ 618 W.m^{-2} \quad [5.263]$$

$$79618 = \lambda_{meas2}\frac{170.8}{0.1} \quad [5.264]$$

We will therefore measure the value of λ_{meas2}:

$$\lambda_{meas2} = 46.61 W.m^{-1}.K^{-1} \quad [5.265]$$

3) In radiation, the temperatures should be systematically expressed in Kelvin:

$$T_a = 19 + 273 = 292\,K \quad [5.266]$$

a) The radiator has a surface

$$S_{rad} = \pi \frac{\left(50.10^{-2}\right)^2}{4} = 0.196 m^2 \qquad [5.267]$$

The flux density coming from the resistance is therefore

$$\varphi = \frac{25}{0.196} = 127.5 W.m^{-2} \qquad [5.268]$$

The thermal balance of the radiator will be written in terms of flux densities. Considering the flux provided by the resistance, the flux radiated by the radiator, the convection flux and the part absorbed from the flux emitted by the atmosphere, assimilated to a black body with a temperature T_a, we will have:

$$\varphi + \varepsilon\sigma T_a^4 - \varepsilon\sigma T_R^4 - h\left(T_R - T_a\right) = 0 \qquad [5.269]$$

$$\varepsilon\sigma T_R^4 \; hT_R - \varphi - \varepsilon\sigma T_a^4 - hT_a = 0 \qquad [5.270]$$

$$5.1.10^{-8}\, T_R^4 + 5T_R - 127.5 - 370.8 - 1460 = 5.1.10^{-8}\, T_R^4 + 5T_R - 1956.5 = 0 \qquad [5.271]$$

A solver gives us T_R

$$T_R = 304.1K = 31.1°\mathrm{C} \qquad [5.272]$$

which is an error on T_R:

$$T_R - T_a = 12.1°\mathrm{C} \qquad [5.273]$$

b) The apparent resistance R_{app} is therefore equal to

$$R_{app} = \frac{T_R - T_a}{\varphi_0} = 1.52.10^{-4}\, W^{-1}.m^2.K \qquad [5.274]$$

c) By applying the approximation

$$T_w^4 - T_a^4 \approx 4T_a^3 (T_W - T_a), \quad [5.275]$$

we can rewrite the above equation in the form:

$$\varphi + \varepsilon\sigma T_a^4 - \varepsilon\sigma T_R^4 - h(T_R - T_a) = \varphi + 4\varepsilon\sigma T_a^3 (T_R - T_a) - h(T_R - T_a) \quad [5.276]$$

$$= \varphi - (4\varepsilon\sigma T_a^3 + h)(T_R - T_a) = 0 \quad [5.277]$$

We have

$$S\varphi_0 = S_{Rad}\,\varphi \quad [5.278]$$

so:

$$\varphi = \varphi_0 \frac{S}{S_{rad}} = \varphi_0 \frac{3.14.10^{-4}}{0.196} = 1.6.10^{-3}\,\varphi_0 \quad [5.279]$$

$$(4\varepsilon\sigma T_a^3 + h)(T_R - T_a) = 1.6.10^{-3}\,\varphi_0 \quad [5.280]$$

$$\frac{(4\varepsilon\sigma T_a^3 + h)}{1.6.10^{-3}}(T_R - T_a) = \varphi_0 \quad [5.281]$$

which is

$$R_{thapp} = \frac{1.6.10^{-3}}{4\varepsilon\sigma T_a^3 + h} \quad [5.282]$$

which gives as a value:

$R_{thapp} = 1.59.10^{-4}$, which is very close to (4.6%) from the result found above.

We note that

$$T_W - T_a = 12.1 \quad [5.283]$$

so

$$\frac{T_W - T_a}{T_a} = 4.1.10^{-2} << 1. \quad [5.284]$$

The approximation therefore leads to an acceptable result.

d) Applying Fourier's law, by recalculating the value of $T_1 - T_2$ while maintaining $T_2 \neq T_a$:

the real difference in temperature between the two faces of the bar in this model is:

$$T_R - T_a = 12.1°\text{C} \quad [5.285]$$

$$T_1 - T_2 = T_1 - T_R = 223 - 19 - 12.1 = 191.9°\text{C} \quad [5.286]$$

$$\varphi_0 = \lambda_{meas3} \frac{T_1 - T_2}{L} = 79618 W.m^{-2} \quad [5.287]$$

$$79618 = \lambda_{meas3} \frac{191.9}{0.1} \quad [5.288]$$

We will therefore measure the value for λ_{meas3}

$$\lambda_{mes3} = 41.5 W.m^{-1}.K^{-1} \quad [5.289]$$

4) Considering all the sources of error

Applying Fourier's law, recalculating the "exact" value of $T_1 - T_2$:

– the real difference in temperature between the two faces of the bar is in this model:

$$T_R - T_a = 12.1°\text{C} \quad [5.290]$$

$$T_1 - T_2 = T_1 - T_R = 223 - 19 - 2*16.6 - 12.1 = 168.7°\text{C} \quad [5.291]$$

$$\varphi_0 = \lambda_{real} \frac{T_1 - T_2}{L} = 79618 W.m^{-2} \quad [5.292]$$

$$79618 = \lambda_{real} \frac{168.7}{0.1} \quad [5.293]$$

We will therefore measure for λ_{meas3} the value:

$$\lambda_{real} = 47.2 W.m^{-1}.K^{-1} \quad [5.294]$$

5) The relative errors committed will therefore be as follows:

By taking as a real value of the thermal conductibility $\lambda_{real} = 47.2 W.m^{-1}.K^{-1}$

Using the measurement without caution

$$\lambda_{mes1} = 37.2 W.m^{-1}.K^{-1} \quad [5.295]$$

The relative error amounts to:

$$\frac{\lambda_{meas1} - \lambda_{real}}{\lambda_{real}} = \frac{47.2 - 37.2}{47.2} = 47\% \quad [5.296]$$

Using the measurement while considering only the air gaps

$$\lambda_{mes2} = 46.61 W.m^{-1}.K^{-1} \quad [5.297]$$

The relative error amounts to:

$$\frac{\lambda_{meas1} - \lambda_{real}}{\lambda_{real}} = \frac{47.2 - 46.61}{47.2} = 12.5\% \quad [5.298]$$

Using the measurement while considering only transfers in the radiator

$$\lambda_{mes3} = 41.5 W.m^{-1}.K^{-1} \quad [5.299]$$

The relative error amounts to:

$$\frac{\lambda_{meas1}-\lambda_{real}}{\lambda_{real}}=\frac{47.2-4.5}{47.2}=12.1\% \qquad [5.300]$$

6) Here, we should solve an unstable conduction problem at an imposed flux density. We seek the evolution of the surface temperature where the flux is imposed.

This will be the classic solution:

$$\theta=T(x,t)-T_i \qquad [5.301]$$

$$\theta=\frac{2\varphi_0}{\lambda}\sqrt{at}\,ierfc(\xi) \qquad [5.302]$$

with:

$$\xi=\frac{x}{\sqrt{4at}} \qquad [5.303]$$

at the surface, we have:

$$\xi=\frac{0}{\sqrt{4at}}=0 \qquad [5.304]$$

$$ierfc(0)=\frac{1}{\sqrt{\pi}}=0.5642 \qquad [5.305]$$

Calculate

$$a=\frac{\lambda}{\rho c}=\frac{47}{7833*465}=1.29.10^{-5}\,m^2.s^{-1} \qquad [5.306]$$

We want the input temperature of the material to be equal to $T_1=233°C$ at almost 1%, which is:

$$T(t)=0.99*233=230.67 \qquad [5.307]$$

The initial temperature will be taken at ambient temperature T_a = 19°C

$$\theta(t) = 230.67 - 19 = 211.67 \quad [5.308]$$

Applying the relationship recalled previously:

$$\theta = \frac{2\varphi_0}{\lambda}\sqrt{at}\, ierfc(\xi) \quad [5.309]$$

$$211.67 = \frac{2*79620}{47}\sqrt{1.29.10^{-5}*t}\ 0.5642 \quad [5.310]$$

which is the response time at 1%:

$$\sqrt{t} = \frac{211.67*47}{2*79620*0.5642*\sqrt{1.29.10^{-5}}} = 30.83 \quad [5.311]$$

$$t = 950s = 15mn50s \quad [5.312]$$

5.5. General problems

In this last section, we will suggest examples of integrating radiation into more general spectral problems.

EXAMPLE 5.22.– Fahrenheit 451. Episode 1

Ray Bradbury named his famous science fiction novel, *Fahrenheit 451*, after the temperature at which paper burns, which expressed in degrees Fahrenheit is $T_{\inf} = 451°F$.

A flat, circular heating plate of diameter $d = 30cm$ dissipates upwards a power $P = 500W$.

We place a sheet of paper on it. Before being placed, the sheet is at the ambient temperature of the room, which is $T_a = 21°C$.

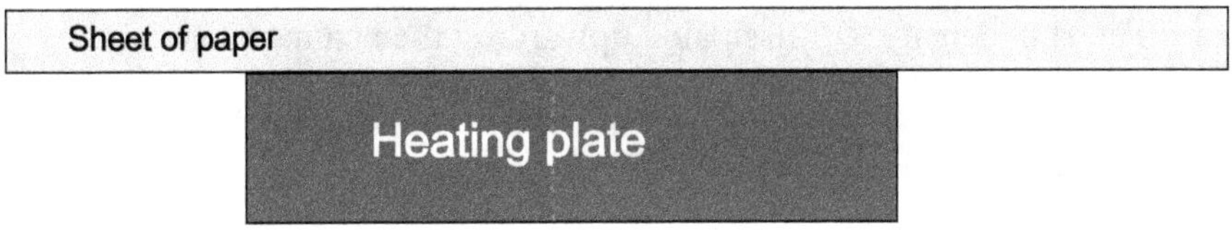

Figure 5.7. *Fahrenheit 451, I*

We give the following properties of the paper:

– thermal conductibility $\lambda = 0.048 W.m^{-2}.K^{-1}$

– density $\rho = 80\,kg.m^{-3}$

– specific heat capacity $c = 2000\,J.kg^{-1}$.

We recall that a temperature expressed in Celsius $T_{celsius}$ can be expressed in Fahrenheit, which is $T_{Farenheit}$ from the formula:

$$T_{Farenheit} = 1.8 T_{Celsius} + 32$$

1) What is the flux density φ_0 dissipated upwards by the heating plate?

2) How long will it take for the paper to burn?

SOLUTION.– The thermal diffusivity of the paper is:

$$a = \frac{\lambda}{\rho c} = \frac{0.048}{80 * 2000} = 3.10^{-7}\, m^2.s^{-1} \qquad [5.313]$$

The burning temperature of the paper expressed in Celsius is such that:

$$451 = 1.8 T_{Celsius} + 32 \qquad [5.314]$$

which is:

$$T_{\inf} = 233°C \qquad [5.315]$$

1) The flux density φ_0 dissipated upward by the heating plate is:

$$S = \pi \frac{d^2}{4} = 7.07.10^{-2}\, m^2 \qquad [5.316]$$

$$\varphi_0 = \frac{P}{S} = \frac{500}{7.07.10^{-2}} \qquad [5.317]$$

$$\varphi_0 = 7072\, W.m^{-2} \qquad [5.318]$$

The increase in temperature of the paper at a time t is given by:

$$T(x,t) - T_a = \frac{2\varphi_0}{\lambda}\sqrt{at}\; ierfc\,\xi \qquad [5.319]$$

$$\xi = \frac{x}{\sqrt{4at}} \qquad [5.320]$$

For the lower side of the paper, $x = 0$, $\xi = 0$ and $ierfc(0) = 0.5642$

The paper will burn when the lower part of the paper is at $T_{Celsius} = 233°C$, which is at

$$T(x,t) - T_a = 233 - 21 = 212°C. \qquad [5.321]$$

This happens for $x = 0$, which is for $\xi = 0$.

The burning time t_{inf} is deduced from this by:

$$T(0,t_{\text{inf}}) - T_a = \frac{2\varphi_0}{\lambda}\sqrt{a t_{\text{inf}}}\; ierfc(0) \qquad [5.322]$$

$$212 = \frac{2*7072}{0.048}\sqrt{3.10^{-7}\, t_{\text{inf}}}\; ierfc(0) = \frac{2*7072}{0.048}\sqrt{3.10^{-7}\, t_{\text{inf}}}\; 0.5642 \qquad [5.323]$$
$$= 91.06\sqrt{t_{\text{inf}}}$$

$$\sqrt{t_{\text{inf}}} = \frac{212}{91.06} = 2.328 \qquad [5.324]$$

$$t_{\text{inf}} = 5.42\,s \qquad [5.325]$$

EXAMPLE 5.23.– Fahrenheit 451. Episode 2

The previous experiment was carried out on yellow paper, considered as a gray body with emissivity $\varepsilon = 0.72$.

We now try to ignite the paper by lighting it with a lamp that gives an illumination $E\,W.m^{-2}$. We place ourselves at the moment when the equilibrium surface temperature is reached.

The ambient temperature is still $T_a = 21°\text{C}$ and the paper always burns at $T_{\text{inf}} = 451°\,F$.

We estimate that the upper surface of the paper is subject to a convection coefficient $h = 5W.m^{-2}.K^{-1}$.

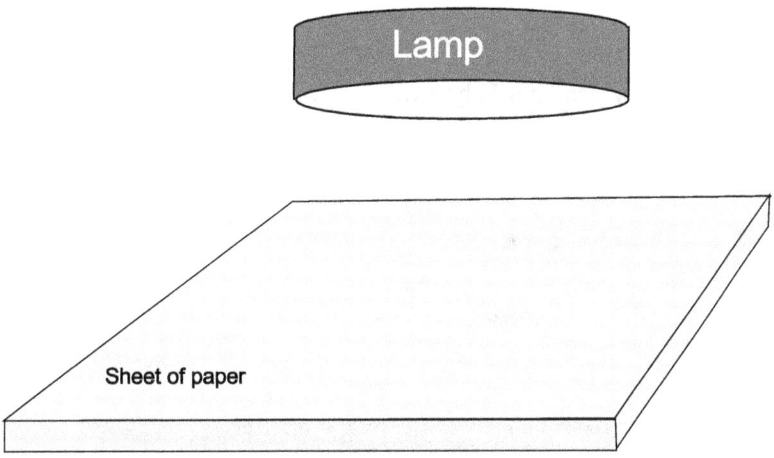

Figure 5.8. *Fahrenheit 451, II*

1) This question should be solved literally.

We will write the temperature reached by the surface of the paper T_W depending on the incident radiation.

Write the thermal balance at the upper surface of the paper.

2) Digital application

a) At what minimum illumination will the paper ignite?

b) What should the maximum diameter of a lamp with power $P = 500W$ be in order to light the paper?

SOLUTION.–

1) The balance expresses that the flux density absorbed is equal to the heat re-emitted:

$$\varepsilon E = \varepsilon \sigma \left(T_W^4 - T_a^4\right) + h\left(T_W - T_a\right) \quad [5.326]$$

2) We have

$T_{\text{inf}} = 233°\text{C}$ which should be expressed in Kelvin:

$$T_{\text{inf}} = 233 + 273 = 506\,K \quad [5.327]$$

$$T_a = 21 + 273 = 294\,K \quad [5.328]$$

a) The balance makes it possible to directly calculate E $\varepsilon E = \varepsilon \sigma \left(T_W^4 - T_a^4\right) + h\left(T_W - T_a\right)$

$$= 0.72 * 5.67.10^{-8}\left(506_W^4 - 294_a^4\right) + 5\left(506 - 294\right) \quad [5.329]$$

$$E = \frac{2371 + 1060}{0.72} \quad [5.330]$$

$$E = \frac{3431}{0.72} = 4765 W.m^{-2} \quad [5.331]$$

b) If $S = \pi \frac{D^2}{4}$ [5.332]

is the surface of the lamp of diameter D, we will have:

$$P = E\,S \quad [5.333]$$

So:

$$S = \frac{P}{E} = \frac{500}{47651} = 0.1049\,m^2 \quad [5.334]$$

And:

$$D = \sqrt{\frac{4S}{\pi}} = \sqrt{\frac{4 * 0.105 S}{\pi}} \quad [5.335]$$

$$D = 0.365\,m = 36.5\,cm \quad [5.336]$$

EXAMPLE 5.24.– Fusion of a steel ingot in an oven

This problem will be solved literally up to and inclusive of question 3. We will only use numerical values for question 4.

A steel ingot is placed, in view of its fusion, in an oven assimilated to a black body whose temperature is $T_F = 1600°C$. The temperature at which the steel fuses is $T_{fusion} = 1430°C$.

Throughout the problem, we will assume that the temperature of the ingot is uniform and equal to its superficial temperature $T_W(t)$.

The steel ingot takes the form of a parallelepiped of dimensions $1m * 0.5m * 0.5m$. The density of the steel is $\rho = 7833 kg.m^{-3}$, and its calorific mass capacity is $c = 465\,J.kg^{-1}$.

The emissivity of the steel, considered as a gray body, is $\varepsilon = 0.5$.

As it is placed in the oven, the ingot can only receive radiation by its upper surface of dimensions $1\,m * 0.5\,m$.

The problem starts at the moment when the **ingot has already reached the temperature of** $T_{FW0} = 1400°C$.

1) Write the expression of the radiation flux contributing to heating the ingot (balance between the radiation absorbed and the radiation emitted by the ingot).

2) Thermal balance

a) Write the thermal balance of the ingot and from this deduce the differential equation that the temperature $T_W(t)$ of the ingot meets.

b) Simplify the writing of this equation in view of solving it. To do this:

we will apply the approximation:

$$T_F^4 - T_W^4 = 4T_F^3 \left(T_F - T_W\right) \qquad [5.337]$$

we will define the variable $\theta(t)$ by

$$\theta(t) = T_F - T_W(t) \qquad [5.338]$$

we will also pose:

$$\theta_0 = T_F - T_{W0} \qquad [5.339]$$

3) Solve the differential equation and give the expression of $\theta(t)$.

4) Starting from the temperature T_{W0}, how long will it take for the ingot to start to melt?

SOLUTION.–

1) The oven behaves as a black body, which sends a flux to the surface S

$$\Phi = S\,\sigma\,T_F^4 \tag{5.340}$$

The part of this flux absorbed by the ingot, a gray body whose absorption coefficient is equal to the emissivity ε

$$\Phi = S\,\varepsilon\sigma\,T_F^4 \tag{5.341}$$

the ingot, a gray body of emissivity ε emits the flux

$$\Phi = S\,\varepsilon\sigma\,T_W^4 \tag{5.342}$$

This results in the energy balance that participates in heating the ingot

$$\Phi = S\,\varepsilon\sigma\left(T_F^4 - T_W^4\right) \tag{5.343}$$

2) Thermal balance

a) During the time dt, the ingot receives the energy that contributes to its heating, which is the balance:

$$\rho C V\,dT_w = S\,\varepsilon\,\sigma\left(T_F^4 - T_W^4\right)dt \tag{5.344}$$

which leads to the differential equation:

$$\rho C V\frac{dT_w}{dt} = S\,\varepsilon\,\sigma\left(T_F^4 - T_W^4\right) \tag{5.345}$$

b) This differential equation, using $T_F^4 - T_W^4 = 4T_F^3\left(T_F - T_W\right)$ is simplified into:

$$-\rho C V\frac{d\theta_w}{dt} = S\,\varepsilon\,\sigma\,4T_F^3\left(T_F - T_W\right) = S\,\varepsilon\,\sigma\,4T_F^3\,\theta \tag{5.346}$$

3) The solution of the equation is classic:

$$-\rho C V \frac{d\theta_w}{dt} = \quad S\varepsilon\sigma\, 4T_F^3\,\theta \tag{5.347a}$$

$$\frac{d\theta_w}{dt} = \quad -\alpha\theta_w \tag{5.347b}$$

α is defined by $\alpha = \dfrac{S\varepsilon\sigma\, 4T_F^3}{\rho C V}$ [5.348]

The equation is written as:

$$\frac{d\theta_w}{\theta_w} = -\alpha\, dt \tag{5.349}$$

$$d\, Ln\theta_w = -\alpha t + Ln\, C \tag{5.350}$$

$$\theta = C \exp{-\alpha t} \tag{5.351}$$

with the boundary condition:

$$\theta = \theta_0 \exp{-\alpha t} \tag{5.352}$$

4) The time can be deduced from the expression of $\theta(t)$:

$$t = -\frac{1}{\alpha} Ln \frac{\theta}{\theta_0} \tag{5.353}$$

α amounts to $\alpha = 4.091.10^{-4}\, s^{-1}$ [5.354]

$$t = -\frac{1}{4.091.10^{-4}} Ln \frac{170}{200} \tag{5.355}$$

$$t = 397.3 s = 6.62 mn \tag{5.356}$$

EXAMPLE 5.25.– Ironing day

NOTE.– *Questions 3 and 4 from the conduction domain are entirely independent of question 2.*

The combination of questions 1 and 2 therefore forms a problem. The combination of questions 1, 3 and 4 therefore forms another problem.

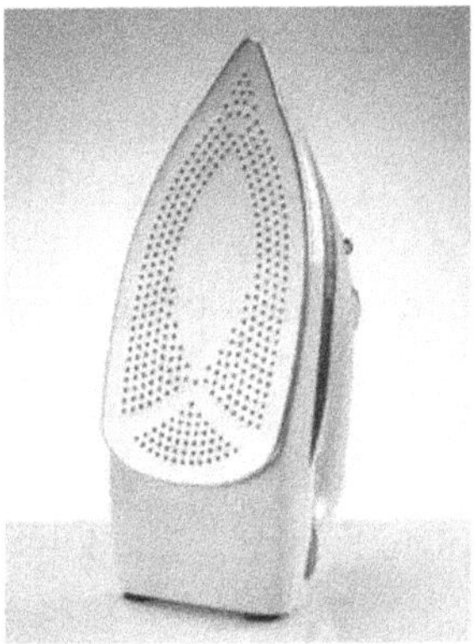

A housewife is ironing her laundry. The iron gives out a power $P = 400W$ through the surface of its soleplate[3] (the surface in contact with the laundry), which has an area of $S = 150\,\text{cm}^2$. The ambient temperature is $T_a = 20°C$. We consider that the atmosphere radiates like a black body.

The iron is placed on a support so that its soleplate is exposed freely to the atmosphere.

We assume throughout the problem that the heat produced by the iron's resistance leaves exclusively and uniformly on the surface of the soleplate.

1) What is the flux Φ that leaves through the soleplate?

What is the flux density φ?

3 For those who do not iron very often, the lower part of the iron, intended to lie on the laundry when it is ironed, is called the soleplate.

2) Two heat transfer processes initially make it possible to produce this flux Φ:

– The thermal convection (here it will be a natural convection). We estimate that the convection coefficient is $h = 120 W\, m^{-2}\, K^{-1}$;

– The radiation. We believe that the iron's soleplate is a gray body with emissivity $\varepsilon = 0.5$.

a) We assume at first that most of the transfer occurs via convection. What then will be the temperature T_{w1} of the soleplate's surface?

b) We assume in a second instance that the radiation forms most of the heat transfer.

i) Write the thermal balance that leads to an equation in which T_{w2}, the temperature of the soleplate for this second hypothesis, is the unknown.

ii) Solve this equation

c) Using the calculations above, which of the two hypotheses, of an exclusively convective or exclusively radiative transfer, seems least unacceptable?

d) A quick question that barely involves calculation. Do you think that, if we wish to consider two modes of transfer, we will find the right solution by accepting the usual approximation $T_{w2}^4 - T_a^4 = 4T_a^3 \left(T_{w2} - T_a\right)$?

3) Since ironing is not a particularly exciting activity, the housewife watched television as she does it.

She accidentally touches the soleplate with her finger.

We will assume that the fingertip has the following characteristics:

– thermal conductivity: $\lambda = 0.2 Wm^{-2} K^{-2}$

– density: $\rho = 900\, kg.m^{-3}$

– specific heat capacity: $c = 4000\, J\, kg^{-1}\, K^{-1}$

We estimate that the initial temperature T_i of the housewife's fingers is 32°C.

a) Taking account of the flux density φ determined in question 1, after how much time t_1 will the temperature of the housewife's fingertip reach 50°C?

b) Taking account of this same flux density φ determined in question 1, and assuming that the housewife keeps her finger on the iron, after how much time t_2 will the temperature of the housewife's fingertip reach 70°C?

4) Sensations are carried from the finger to the brain through nerve impulses that propagate at a finite speed. We will admit here that a delay of $\tau = 40\,ms$ is needed so that pain can be transmitted from the finger to the brain.

We believe that at 50°C, a sensation of pain appears. The housewife will therefore feel pain 40 ms after the temperature of her finger has reached 50°C.

We consider that a severe second-degree burn is likely as soon as the temperature of the finger reaches 70°C.

Taking account of the calculations in question 2, and supposing that the woman is not entirely engrossed in the TV adverts, will the housewife withdraw her finger before sustaining a serious burn? Will she need good reflexes?

SOLUTION.–

1) The flux Φ put out through the soleplate is equal to the power of the iron:

$$\Phi = P = 400W \qquad [5.356]$$

$$\varphi = \frac{400}{150.10^{-4}} = 2.67.10^{4}\,Wm^{-2} \qquad [5.357a]$$

2) Two heat transfer processes make it possible at first to provide this flux Φ: thermal convection and radiation.

a) Transfer via thermal convection

$$\varphi = h\left(T_{w1} - T_a\right) \tag{5.357b}$$

$$2.67.10^4 = 120 * \left(T_{w1} - 20\right) \tag{5.358}$$

$$T_{w1} = 242.5°\mathrm{C} \tag{5.359}$$

b) Transfer by radiation

i) The radiation balance at the surface is calculated from an expression analyzed in **Chapter 10**:

$$\varphi = \varepsilon\sigma\left(T_{w2}^4 - T_a^4\right) \tag{5.360}$$

ii) An equation which is solved by:

$$\varphi = \varepsilon\sigma\left(T_{w2}^4 - T_a^4\right) \tag{5.361}$$

$$2.67.10^4 = 0.5 * 5.67.10^{-8} * \left(T_{w2}^4 - 293^4\right) \tag{5.362}$$

$$T_{w2}^4 - 7.37.10^9 = 9.418.10^{11} \tag{5.363}$$

$$T_{w2} = 987\,K = 714°\mathrm{C} \tag{5.364}$$

c) The convective hypothesis leads to a lower temperature. Convection is therefore dominant.

d) What do we say of the usual approximation $T_{w2}^4 - T_a^4 = 4T_a^3\left(T_{w2} - T_a\right)$?

Note that $T_{w2} - T_a = 242.5 - 20 = 222.5$ is a long way from being much smaller than $T_a = 293$!

This will not work.

3) The housewife accidentally touches the iron's soleplate with her finger.

a) We know the flux density φ determined in question 1

$$\varphi_0 = 2.67.10^4\, Wm^{-2} \quad [5.365]$$

$$\theta = \theta\left(\varphi_0, t, \xi\right) \quad [5.366]$$

To calculate the time t_1 needed for the temperature of the housewife's fingertip to reach 50°C, we need to solve an imposed flux density problem φ_0

$$\theta = T(x,t) - T_i \quad [5.367]$$

$$\theta = \frac{2\varphi_0}{\lambda}\sqrt{at}\; ierfc(\xi) \quad [5.368]$$

At the surface of the finger

$$x = 0 \;;\; \xi = \frac{x}{2\sqrt{at}} = 0 \quad [5.369]$$

$$ierfc\,(0) = 0.5642 \quad [5.370]$$

$$\theta = T(x,t) - T_i \quad [5.371]$$

$$\theta = T(0,t) - T_i = \frac{2\varphi_0}{\lambda}\sqrt{at}\; ierfc(0) \quad [5.372]$$

$$\theta = T(\,0,t) - 32 = \frac{2 * 2.67.10^4}{0.2}\sqrt{5.55.10^{-8} * t_1} * 0.5642 \quad [5.373]$$

$$\theta = 50 - 32 = \frac{2 * 2.67.10^4}{0.2}\sqrt{5.55.10^{-8} * t_1} * 0.5642 \quad [5.374]$$

$$18 = 35.5\sqrt{t_1} \qquad [5.375]$$

$$t = 0.26\ s \qquad [5.376]$$

b) Assuming that the housewife keeps her finger on the iron, calculate the time t_2 for the housewife's fingertip to reach 70°C.

$$x = 0\ ;\ \xi = \frac{x}{2\sqrt{at}} = 0 \qquad [5.377]$$

$$ierfc\left(0\right) = 0.5642 \qquad [5.378]$$

$$\theta = T\left(x,t\right) - T_i \qquad [5.379]$$

$$\theta = T\left(0,t\right) - T_i = \frac{2\varphi_0}{\lambda}\sqrt{at}\ ierfc\left(0\right) \qquad [5.380]$$

$$\theta = T\left(0,t\right) - 32 = \frac{2 * 2.67.10^4}{0.2}\sqrt{5.55.10^{-8} * t_2} * 0.5642 \qquad [5.381]$$

$$\theta = 50 - 32 = \frac{2 * 2.67.10^4}{0.2}\sqrt{5.55.10^{-8} * t_2} * 0.5642 \qquad [5.382]$$

$$38 = 35.5\sqrt{t_2} \qquad [5.383]$$

$$t_2 = 1.14\ s \qquad [5.384]$$

4) Sensation of burning and actually burning the finger

The sensation of pain will reach the brain after

$$t_D = 0.26 + 0.04 = 0.3s < 1.14s \qquad [5.385]$$

This works, but only just; she will need good reflexes.

EXAMPLE 5.26.– The thermal problem of a wooden construction in a very sunny location

A wooden garage painted dark gray is exposed to full sun one hot summer day. The owner of this building (also the author of the present problem) knows that this paint flakes and falls apart. Intrigued by the high temperature of the wall, he decides to make an estimate of this temperature that will be denoted by T_W. He considers the following points:

– the solar radiation is $E = 700 W.m^{-1}$ at midday;

– the ambient temperature is $T_a = 25$°C;

– the atmosphere radiates like a black body at temperature T_a;

– the exterior surface of the wooden wall radiates like a gray body, emissivity $\varepsilon = 0.9$. The wood therefore absorbs 90% of the incident radiation, whether it comes from the Sun or from the atmosphere.

The wall has a thickness $e = 10$cm and the wood it is made of has a thermal conductivity $\lambda_B = 0.1 W.m^{-1}.K^{-1}$.

Exchanges via convection are characterized by an exterior convection coefficient $h_e = 10 W.m^{-2}.K^{-1}$ and an interior convection coefficient $h_i = 10 W.m^{-2}.K^{-1}$. We will consider that the thermal exchanges inside the garage are purely convective.

The temperature inside the garage is $T = 20$°C. The temperature of the inside face of the wall is denoted by T_i.

1) Detail the different fluxes to be taken into account to achieve the thermal balance at the exterior surface of the garage.

2) Write the thermal balance on the external face of the wall.

3) Give the value of the exterior temperature of the wall.

We will use two methods for this:

a) an exact resolution of the equation found with recourse to a solver;

b) using approximation

$$T_W^4 - T_a^4 \approx 4T_a^3\left(T_W - T_a\right)$$

the use of which will be justified in this case.

Do you think that an ordinary paint can last a long time?

SOLUTION.–

1) We will write the terms of this balance for a surface unit of the wall:

in terms of radiation:

– part of the solar radiation absorbed by the external face of the wall

$$\Phi_{R1} = \varepsilon E \qquad [5.386]$$

– part of the atmospheric radiation absorbed by the external face of the wall. This radiation is that of a black body at temperature T_a

$$\Phi_{R2} = \varepsilon \sigma T_a^4 \qquad [5.387]$$

radiation admitted by the external face of the wall

$$\Phi_{R3} = \varepsilon \sigma T_W^4 \qquad [5.388]$$

in terms of convection:

– convective exchange between the external face of the wall and the atmosphere

$$\Phi_{C1} = h_e\left(T_W - T_a\right) \qquad [5.389]$$

– exchange between the external face of the wall and the inside of the garage. This transfer links conduction and convection. It will be characterized by a thermal resistance per m².

$$R_{th} = \frac{e}{\lambda} + \frac{1}{h_i} : \quad [5.390]$$

$$\Phi_{C2} = \frac{T_W - T_a}{R_{th}} \quad [5.391]$$

2) The thermal balance written for one meter square of the external face of the wall is therefore written, counting positively the fluxes received by the external face of the wall and negatively the fluxes leaving this wall:

$$\Phi_{R1} + \Phi_{R2} - \Phi_{R3} - \Phi_{C1} - \Phi_{C2} = 0 \quad [5.392]$$

$$\varepsilon E + \varepsilon \sigma T_a^4 - \varepsilon \sigma T_W^4 - h_e \left(T_W - T_a\right) - \frac{T_W - T_a}{R_{th}} = 0 \quad [5.393]$$

which gives a fourth-degree equation in T_W

$$\varepsilon \sigma T_W^4 + \left(h_e + \frac{1}{R_{th}} \right) T_W - \varepsilon E - \varepsilon \sigma T_a^4 - \left(h_e + \frac{1}{R_{th}} \right) T_a = 0 \quad [5.394]$$

We have $R_{th} = \frac{0.1}{0.1} + \frac{1}{10} = 1.1 K.m^{-2}.W^{-1}$ [5.395]

$$5.103.10^{-8} T_W^4 + 10.91 T_W - 630 - 402.4 - 3251 = 0 \quad [5.396]$$

$$5.103.10^{-8} T_W^4 + 10.91 T_W - 4283.4 = 0 \quad [5.397]$$

3) *We can seek* T_W *by solving the equation directly with the aid of a solver.*

We therefore find:

$$T_W = 334.8 K = 61.8°\text{C} \quad [5.398]$$

Otherwise, we can try to use the approximation:

$$T_W^4 - T_a^4 \approx 4T_a^3 \left(T_W - T_a\right) \quad [5.399]$$

And rewrite the equation in approximate form:

$$\varepsilon\sigma\left(T_W^4 - T_a^4\right) + \left(h_e + \frac{1}{R_{th}}\right)\left(T_W - T_a\right) - \varepsilon E = 0 \qquad [5.400]$$

$$\left[4\varepsilon\sigma T_a^3 + \left(h_e + \frac{1}{R_{th}}\right)\right]\left(T_W - T_a\right) - \varepsilon E = 0 \qquad [5.401]$$

$$\left[5.402 + (10.91)\right]\left(T_W - T_a\right) - 630 = 0 \qquad [5.402]$$

$$T_W - T_a = 38.62 \qquad [5.403]$$

$$T_W = 63.69°\text{C} \qquad [5.404]$$

which is close to the exact value at nearly 2.9%

The approximation is only acceptable, in principle, if

$$T_W - T_a << T_W \text{ et } T_W - T_a << T_a \qquad [5.405]$$

In this case, we write:

$$T_W - T_a = \varepsilon T_a \qquad [5.406]$$

and ε *should be "infinitely" small*

We note that here:

$$T_W = 334.8\,K \ ; \ T_a = 298\,K$$

ε amounts to

$$\varepsilon = \frac{T_W - T_a}{T_a} = \frac{334.8 - 298}{298} = 0.123 \qquad [5.407]$$

which justifies the approximation. We understand this better by comparing:

$$(1+\varepsilon)^4 = 1.59 \qquad [5.408a]$$

and

$$1+4\varepsilon = 1.49 \qquad [5.408b]$$

which are equal to nearly 6%.

In any case, a temperature in the order of $T_W = 64°\text{C}$ *can prove decidedly destructive for an ordinary, slightly fragile paint.*

6

Modeling and Simulations under ANSYS

6.1. Conduction, convection and radiation

a) Thermal conduction corresponds to the transport of heat via molecular contact, without displacing matter.

The theoretical mathematical formula that makes it possible to represent conduction is as follows:

$$\phi = (T1 - T2).\, S \,.\frac{\lambda}{e} \qquad [6.1]$$

with:

ϕ = heat flux (in Watts);

λ = thermal conductivity (in Watts/(meters Kelvin));

T1 = “hot” temperature of the system (in°C);

T2 = “cold” temperature of the system (in°C);

e = thickness of the system (in meters);

S = section of the system (in meters squared).

For a color version of all the figures in this chapter, see www.iste.co.uk/ledoux/heat2.zip.

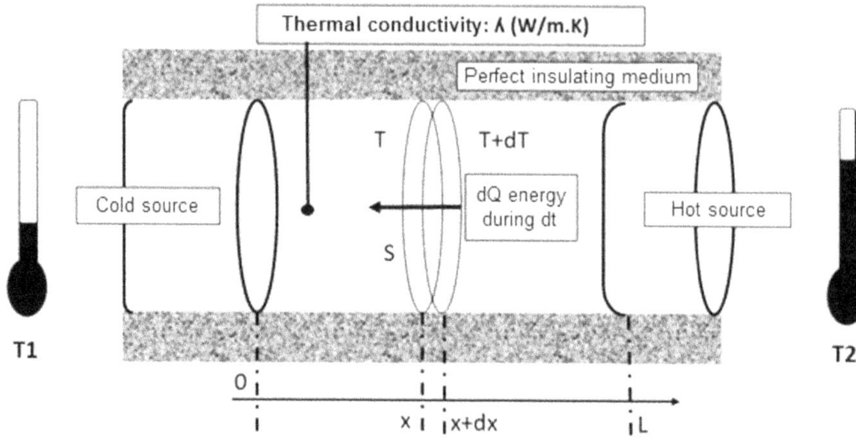

Figure 6.1. *Schematic of conduction in a tube*

b) Convection corresponds to a heat transfer via particle displacement (mass transfer) in a fluid medium (gas or liquid). The characteristic equation is as follows:

$$\phi = (T1 - T2).S.\alpha \qquad [6.2]$$

where:

ϕ = heat flux (in W/m^2);

α = convection coefficient (in W/m^2°C);

T1 = "hot" temperature of the system (in°C);

T2 = "cold" temperature of the system (in°C);

S = surface of the system (wall) (in m^2).

There are two types of convection: natural convection and forced convection. In natural convection, the fluid involved is in motion under the effect of a field of exterior forces (such as gravity), as well as under the effect of densities (due to the differences in temperature on system boundaries).

When the movement of the fluid results from a process independent of differences in the temperature between the walls, the convection is called "forced".

c) Thermal radiation corresponds to heat transfer by sending radiation from the hot body to the cold body. The most concrete example is the Sun's rays on the windows of a house when it causes the temperature in the room to rise.

The characteristic equation expressing the radiation is:

$$\phi = (T1 - T2).S.\rho \qquad [6.3]$$

with:

ϕ = heat flux (in W/m^2);

ρ = radiation coefficient (in $W/m^2°C$);

T1 = "heat" system temperature (in°C);

T2 = "cold" system temperature (in°C);

S = surface of the system (wall) (in m^2).

Figure 6.2 summarizes three types of thermal transfer with a simple example.

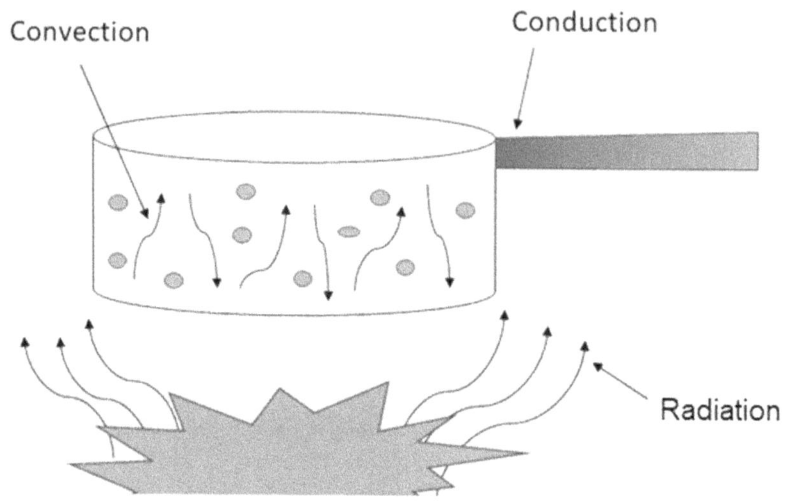

Figure 6.2. *Three types of thermal transfer summarized using a casserole*

6.2. Conduction and convection using ANSYS software

The software uses the first law of thermodynamics, which states that energy is conserved. It is defined using the following equation:

$$\rho c\left(\frac{\partial T}{\partial t}+\{v\}^T\{L\}T\right)+\{L\}^T\{q\}=\ddot{q} \qquad [6.4]$$

where:

ρ = density (DENS in the MP command);

c = specific heat (C in the MP command);

T = temperature;

T = time;

$\ddot{q}$ = *Heat generation rate*;

$\{q\}$ = heat flow vector;

$$\{L\} = \begin{Bmatrix} \dfrac{\partial}{\partial x} \\ \dfrac{\partial}{\partial y} \\ \dfrac{\partial}{\partial z} \end{Bmatrix};$$

$$\{L\} = \begin{Bmatrix} v_x \\ v_y \\ v_z \end{Bmatrix} = \text{ speed vector for heat transport;}$$

$\{L\}T = \nabla T =$ temperature gradient;

$\{L\}^T\{q\} = \nabla.\{q\} =$ divergence of heat flux.

To link the heat flux and the vector flux to the thermal gradient, the software uses Fourier's law:

$$\{q\} = -[D]\{L\}T \qquad [6.5]$$

with $[D] = \begin{bmatrix} K_{xx} & 0 & 0 \\ 0 & K_{yy} & 0 \\ 0 & 0 & K_{zz} \end{bmatrix} =$ conductivity matrix.

By combining different equations, we obtain a more general one:

$$\rho c\left(\frac{\partial T}{\partial t} + v_x\frac{\partial T}{\partial x} + v_y\frac{\partial T}{\partial y} + v_z\frac{\partial T}{\partial z}\right) = \dddot{q} + \frac{\partial}{\partial t}\left(K_x\frac{\partial T}{\partial x}\right) + \frac{\partial}{\partial t}\left(K_y\frac{\partial T}{\partial y}\right) + \frac{\partial}{\partial t}\left(K_z\frac{\partial T}{\partial z}\right) \qquad [6.6]$$

To best define the problem in terms of conduction and convection (mixing the two effects), three main condition limits are imposed:

– a specific temperature T*;

– a specific heat flux q*;

– a specific convective surface (Newton's law of cooling) defined by:

- $h_f (T_s - T_B)$ with
- "coefficient film" = convection coefficient;
- T_B = apparent temperature;
- T_S = temperature of the model's surface.

When initializing the calculation, the software drives the matrices to draw out the terms of convection and conduction.

6.2.1. *Representation of the temperature field*

The temperature in the previous equations can vary in time and space, so the temperature variable T can be divided into the following (differential form):

$$T = \{N\}^T \{T_e\}$$

with:

T = temperature field = T (x, y, z, t);

$\{N\} = \{N(x,y,z)\}$ = form of the elements (interpolation matrix);

$\{T_e\} = \{T_e(t)\}$ = lth vector of nodal temperatures.

The equation used and analyzed by the software is in the discretized form, which can be formulated by:

$$[C_e^t]\{\dot{T}_e\}+([K_e^{tm}]+[K_e^{tb}]+[K_e^{tc}])\{T_e\}=[Q_e^f]+[Q_e^c]+[Q_e^g] \quad [6.7]$$

with:

$[C_e^t]=\rho\int_{vol} c\{N\}\{N\}^T d(vol)$ = thermal capacity matrix;

$\{\boldsymbol{T_e}\}$ = nodal temperature vector;

$[\boldsymbol{K_e^{tm}}] = \rho\int_{vol} c\{N\}\{v\}^T[B]d(vol)$ = conductivity transport matrix for mass elements;

$[K_e^{tb}] = \int_{vol} [B]^T [D][B] d(vol) =$ conductivity matrix for diffusive elements;

$[K_e^{tc}] = \int_{S_3} h_f \{N\}\{N\}^T d(S_3) =$ surface conductivity matrix for convective elements;

$[Q_e^f] = \int_{S_2} \{N\} q * d(S_2) =$ flux vector for mass elements;

$[Q_e^c] = \int_{S_3} c\{N\}\{N\}^T d(vol) =$ vector for thermal surface fluxes from convective elements;

$[Q_e^g] = \int_{vol} \dddot{q}\{N\} d(vol) =$ heat force.

6.3. Radiation using ANSYS software

Radiant energy that transfers between the surfaces of the object of study can have a substantial impact on the problem of thermal transfers.

The following equation is used by the software to visualize the problem of radiation:

$$\sum_{t=1}^{N}\left(\frac{\partial_{ji}}{\epsilon_i} - F_{ji}\frac{1-\epsilon_i}{\epsilon_i}\right)\frac{1}{A_i} Q_i = \sum_{i=1}^{N}(\partial_{ji} - F_{ji})\sigma T_i^4 \qquad [6.8]$$

with:

N = number of radiating surfaces;

δ_{jt} = Kronecker's symbol;

ϵ_i = effective thermal emissivity;

F_{ji} = radiation factor.

The software uses four methods for analyzing radiation:

1) "radiation link element" (LINK31) for simple problems (between two points);

2) "surface effect elements" (SURF151 &152) for radiations between a surface and a point;

3) "radiation matrix method" for general radiation problems involving two or more surfaces;

4) "radiosity solver method" for 3D problems involving two or more surfaces (coupling with the conduction problem).

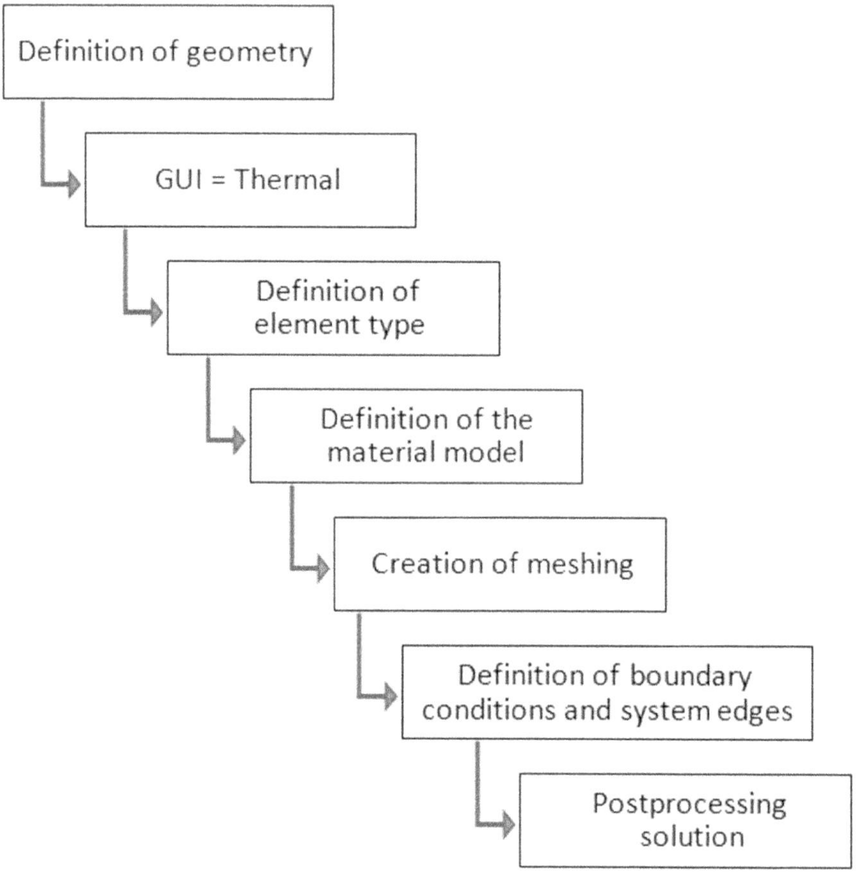

Figure 6.3. *Procedure to use with ANSYS*

6.4. Examples of modeling and analysis with ANSYS

In this example, we will mainly use the ANSYS Mechanical software APDL 16.2.

This is why we will provide some examples of the thermal module in order to practice, before using the real industrial application and the electronic card.

The solutions for the first two examples appear in the stationary state, while the third example is studied in a transitional regime. The first three examples are from the software tutorial by ANSYS "University of Alberta – ANSYS Tutorials". The following examples are taken from the book *Engineering Analysis with ANSYS Software* by T. Stolarski, Y. Nakasone and S. Yoshimoto.

6.4.1. *Simple thermal conduction*

We begin with an example of simple schematized thermal conduction shown in Figure 6.4.

Figure 6.4. *Study model with different boundary conditions*

We therefore developed our model using ANSYS.

We chose the element type "Thermal Mass Solid, Quad 4Node 55" that corresponded to the elements of four nodes with a single degree of freedom (temperature) at each node.

We have also defined the material properties in the form of a thermal conductivity model that is isotropic with a thermal conductivity coefficient k of 10 W/m°C.

The mesh is made up of elements with a surface of 0.05 m^2.

The mesh is illustrated in Figure 6.5.

Figure 6.5. *Model, built and meshed*

We then imposed temperature boundary conditions at the edges of the model:

– 500°C on the upper edge;

– 100°C on the other three edges.

Figure 6.6 shows the model subject to these temperature boundary conditions (represented by yellow triangles).

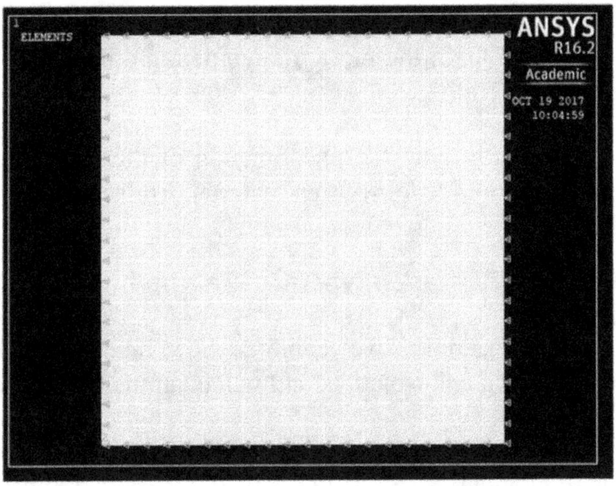

Figure 6.6. *Model under boundary conditions*

After fully defining our model for the example of simple thermal conduction, we obtained the result presented in Figure 6.7.

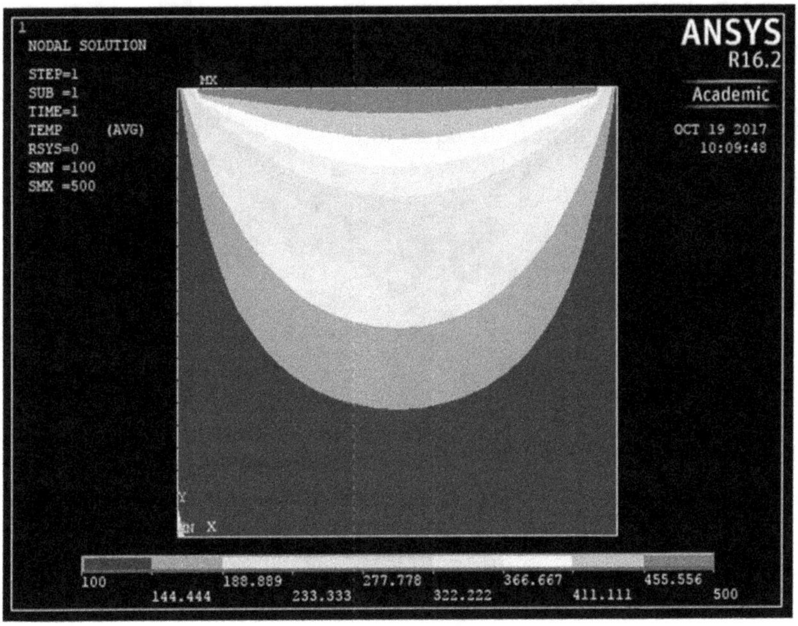

Figure 6.7. *Result from the model showing the effect of simple conduction*

We can see in this example that the more we move away from the plate's upper side, fixed at 500°C, the more the temperature decreases towards the lower temperature of 100°C.

Here, we can clearly note the phenomenon of thermal conduction with heat displacement in our model without material displacement.

6.4.2. *Mixing conduction/convection/isolation*

In the second example, we combine the boundary conditions of conduction, convection and isolation, and will examine the result.

The model used is the same (squared on one side 1 and high on 1), with the same type of element, the same thermal conductivity and the same mesh, but with loads and conditions at different boundaries:

– we impose a temperature of 500°C on the upper side;

– we impose a temperature of 100°C on the left-hand side;

– a boundary condition of convection on the right-hand side with a "coefficient film" of 10 W/m^2°C and an ambient temperature (bulk temperature) of 100°C;

– a boundary condition of isolation on the lower side of the model defined with a "coefficient film" equal to 0.

Figure 6.8. *shows our model in ANSYS, meshed and fully defined.*

Figure 6.9 shows our model in ANSYS, meshed and fully defined.

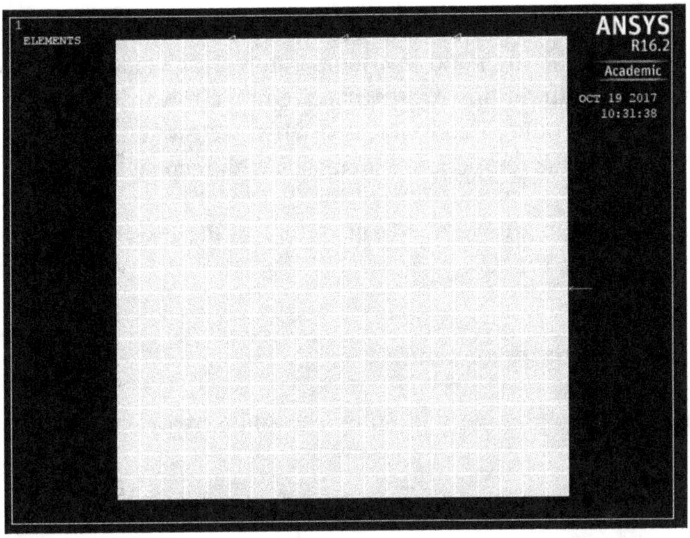

Figure 6.9. *Model using ANSYS with loads and conditions*

The result obtained after the calculation is presented in Figure 6.10.

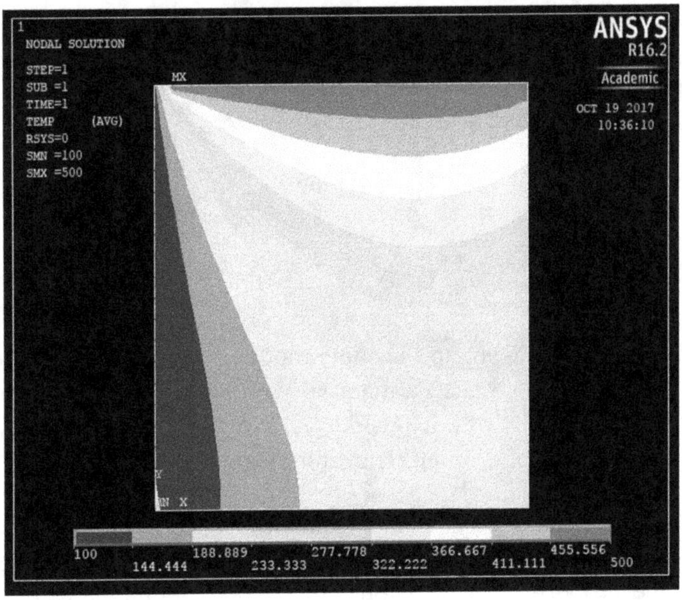

Figure 6.10. *Result of mixing conditions (convection/conduction/isolation)*

When we move away from the high temperature zone, a duality can be observed between the high and low temperatures. The sides imposed remain at the same temperature; however, when we move towards the bottom of the hot zone, the temperature decreases, and when we move away from the cold zone (to the right), the temperature increases with respect to it.

Finally, we obtain an average temperature in the lower right-hand corner of the square (about 250°C).

6.4.3. *Transient thermal conduction*

In the third example, we will study the same model but for a transitory thermal conduction.

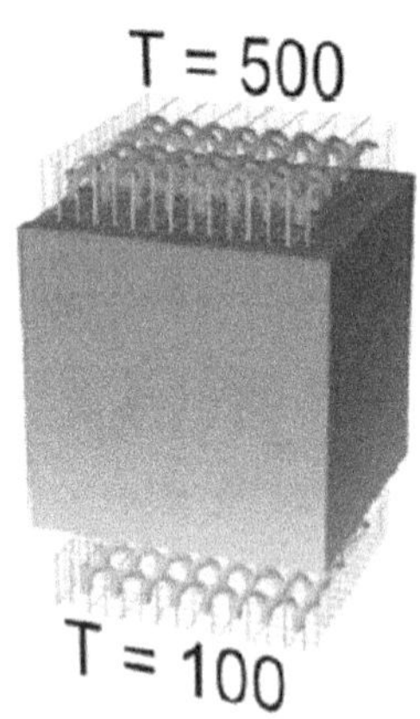

Figure 6.11. *Transistor*

This third example is again the same model, but with different loads and boundary conditions. We used a material with the density 920 kg/m^3 and a specific thermal capacity of 2.040 kJ/kg*K. We no longer seek a steady-state solution, but rather a solution in a transitory state.

From Figures 6.12–6.17, we can therefore observe that the higher temperature (500°C) tends to propagate in the model over time. At a time T=300 s, the proportion of the model that is at a temperature of 500°C is greater than at the start of the test.

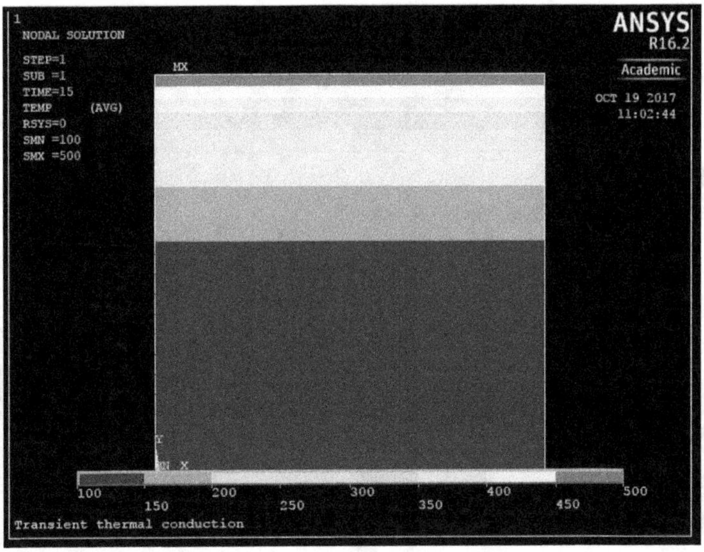

Figure 6.12. *Results of the temperature evolving at the initial state*

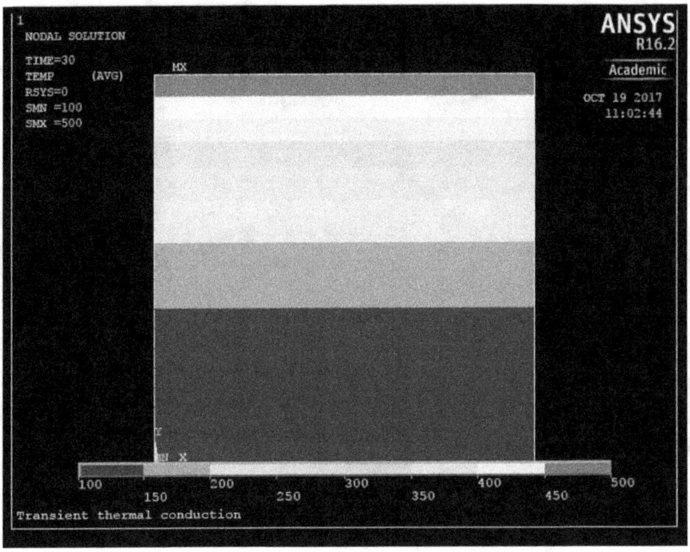

Figure 6.13. *Results of the temperature evolving after 30 seconds*

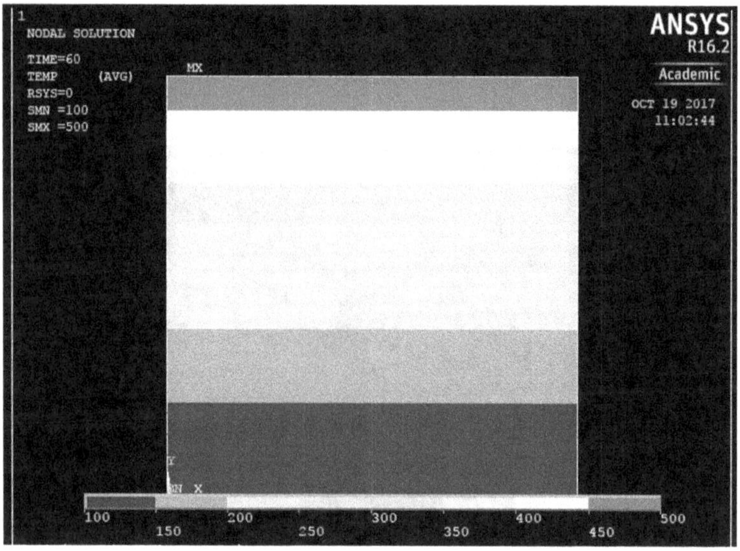

Figure 6.14. *Results of the temperature evolving after 60 seconds*

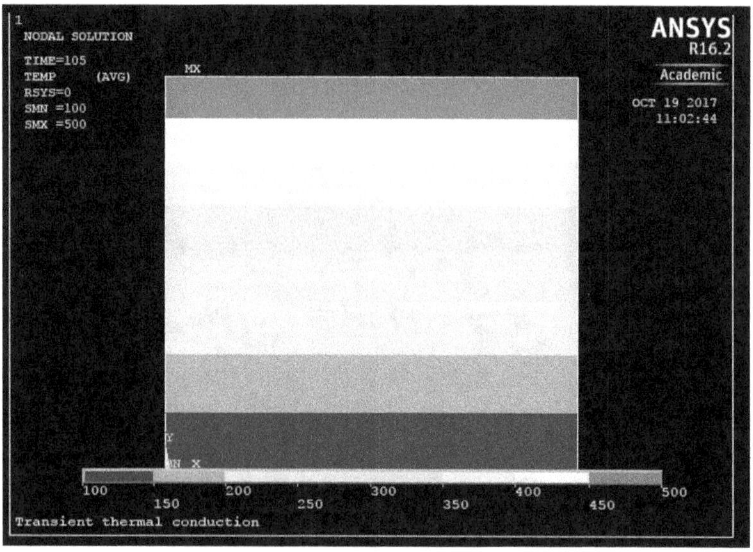

Figure 6.15. *Results of the temperature evolving after 105 seconds*

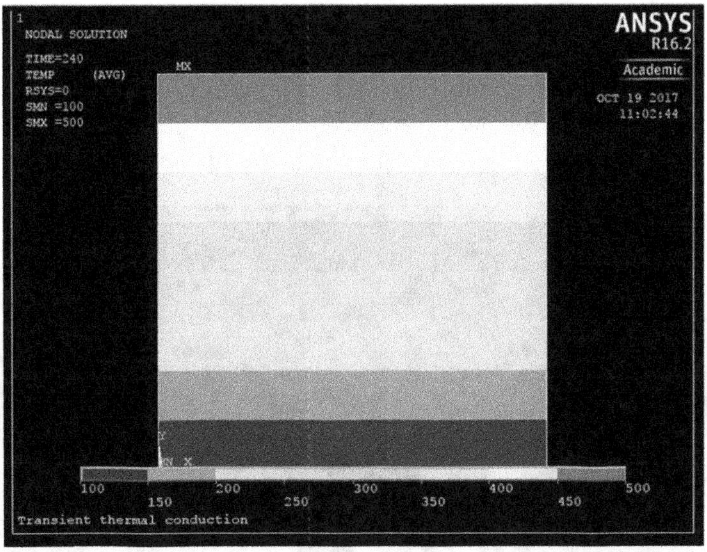

Figure 6.16. *Results of the temperature evolving after 240 seconds*

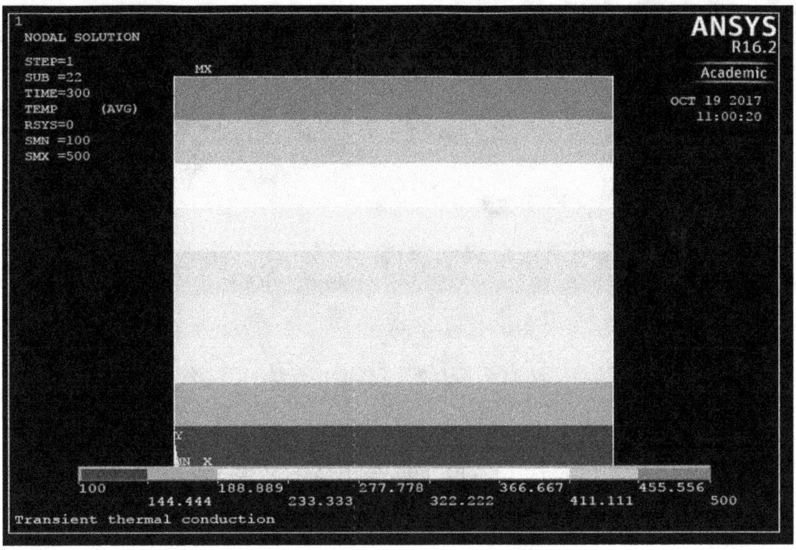

Figure 6.17. *Results of the temperature evolving at the end of our study (T=300 s)*

From the graph shown in Figure 6.18, we can see that for the central node of our mesh and over time, the temperature stabilizes at an average temperature of 300°C. This graph is not entirely finite as it will not completely converge. We need to reduce the size of the mesh or increase the number of time increments in order to obtain the final curve.

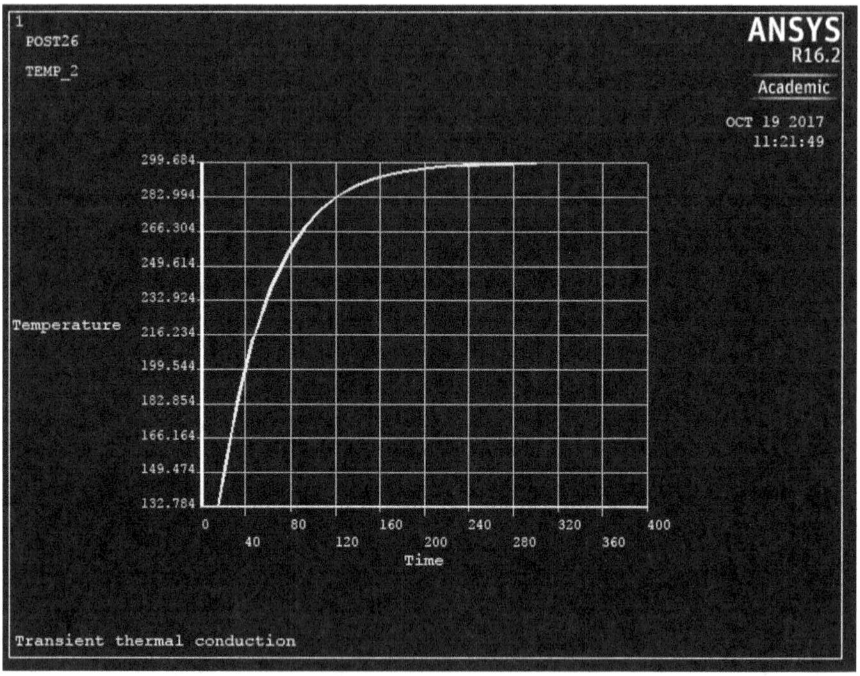

Figure 6.18. *Evolution of the temperature according to time for the central node of our mesh*

6.4.4. *Study of thermal transfers from a brick wall and a cement wall (application to an oven)*

To use the examples developed, I have selected the book entitled *Engineering Analysis with ANSYS Software* by Stalarski, Nakasone and Yoshimoto.

The exercise is a study of thermal transfers crossing two different walls (with different properties):

– the lower part of the cement wall with a thermal conductivity of $k_c = 0.01$ W/mK;

– the external part of the wall is made of brick with a thermal conductivity $k_b = 0.0057$ W/mK.

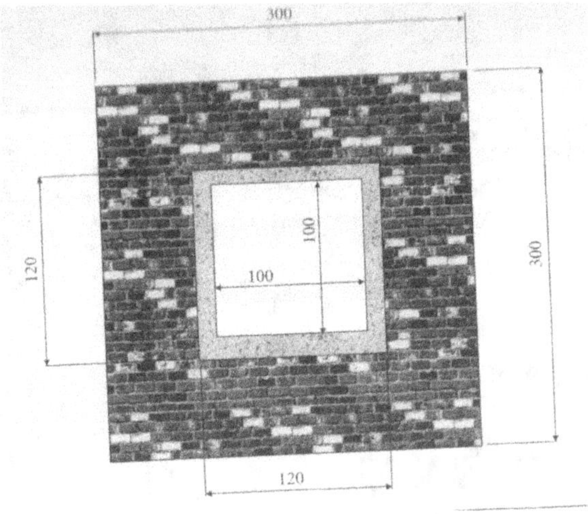

Figure 6.19. *Model for the thermal study*

The lower part is at a temperature of 673 K and has a convective thermal transfer coefficient k_l =0.208 W/m^{2}K. The external part of the furnace is exposed to the ambient air, and is therefore at a temperature of 253 K, with a thermal convection coefficient of $k_2 = 0.068$ W/m^{2}K.

This example is based on a 2D problem, and we will use the graphic interface GUI (graphical user interface in thermics).

The first aim of this example is to determine the temperature distribution in the cement and brick walls in a stationary state. The second aim is to determine the heat exchanges in each wall.

Creating this example has taught us the commands used to create models by subtracting geometries, “picking up” geometries.

For the mesh, we chose the following element type: Thermal Mass Solid, Quad 4node 55 (Reference PLANE 55 in ANSYS). We also defined an element size of 0.5 m.

Figure 6.20 shows the model studied after meshing.

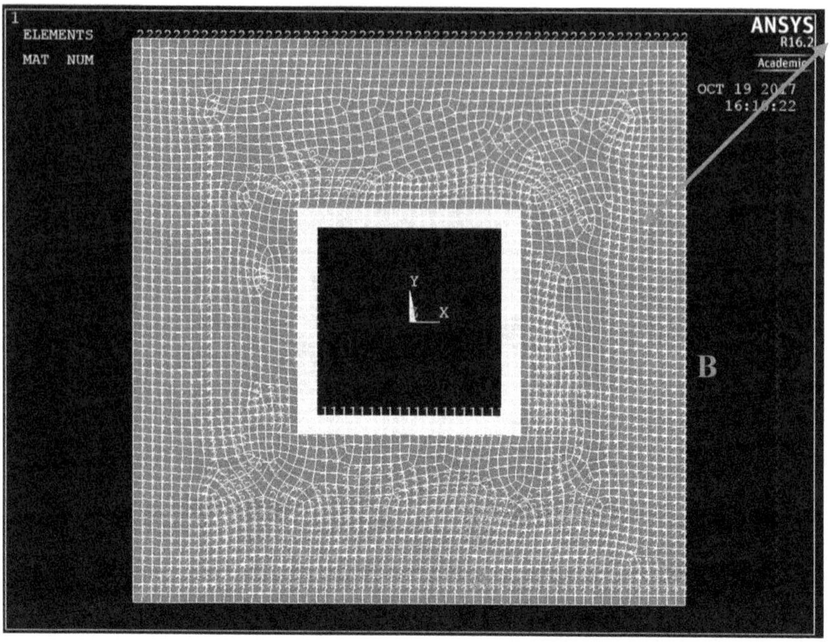

Figure 6.20. *Geometry of our meshed model*

The loads imposed are as follows:

– thermal convection on the model's interior boards

- film coefficient = 0.208 W/m^{2}K;
- bulk temperature = 673 K;

– thermal convection on the model's exterior boards

- film coefficient = 0.068 W/m^{2}K;
- bulk temperature = 253 K.

After defining the loads, we represent them on the graph using the plot–lines function. Figure 6.21 shows the loads imposed on the model in the form of arrows.

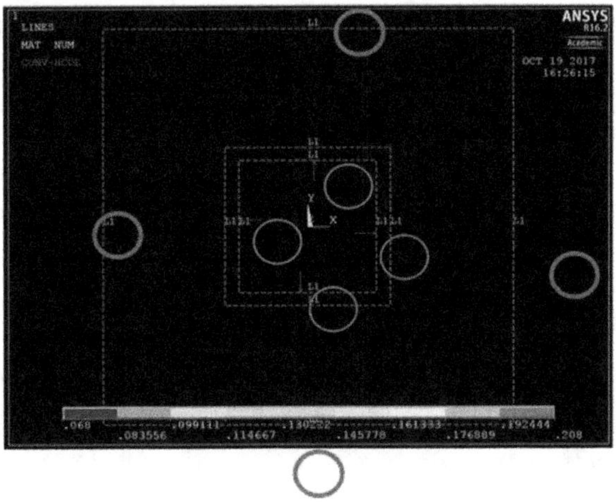

Figure 6.21. *Model under loads*

After defining our whole model, we begin the calculation.

The results are shown below.

First of all, we model the temperature distribution in the model studied (Figure 6.22).

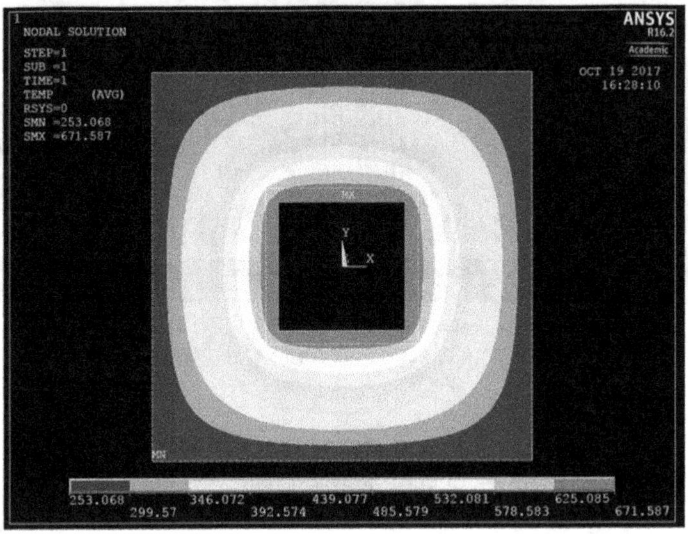

Figure 6.22. *Temperature distribution in the oven*

We can therefore clearly note the phenomenon of convection in our model.

The more we move away from the heat source (the lower part of the model), the more the heat is dissipated. We obtain a significant temperature gradient of the heat source towards the temperature outside the oven (ambient temperature).

Then, we model the thermal flux present in our model oven.

Figure 6.23 shows the heat fluxes present in our model when there is a high temperature imposed inside of the model. The fluxes are directed well from high to low temperatures.

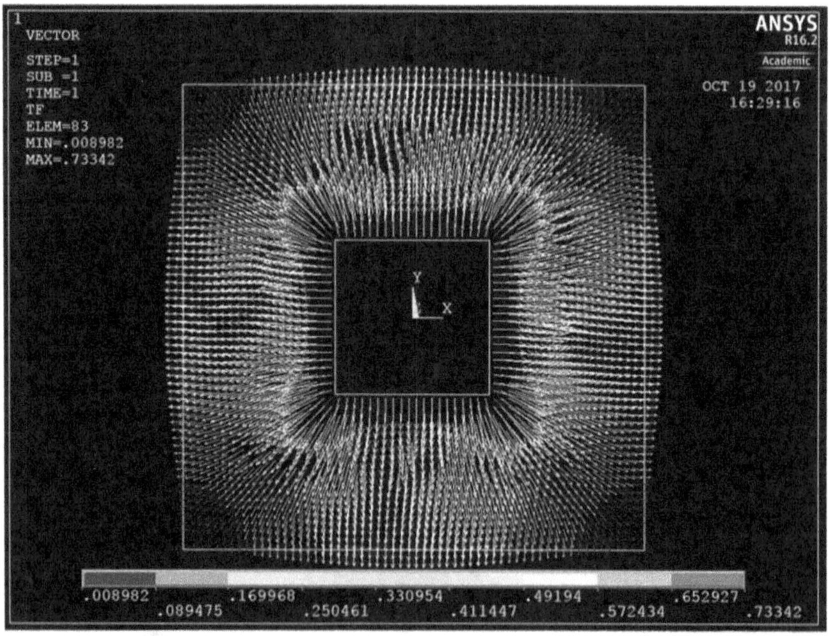

Figure 6.23. *Thermal fluxes present in the walls of the oven*

We note that there is a concentration in the corners inside the oven. This is because the heat is more concentrated at this level.

The last results that we present are the thermal gradients following X (Thermal grad TGX), Y (Thermal grad TGY) and the sum of the two (Thermal grad TGSUM) by considering a segment of our model.

Figure 6.24 summarizes the results of varying the temperature gradient depending on the segment AB. The sum of the gradients decreases markedly when they are removed from the interior of the oven.

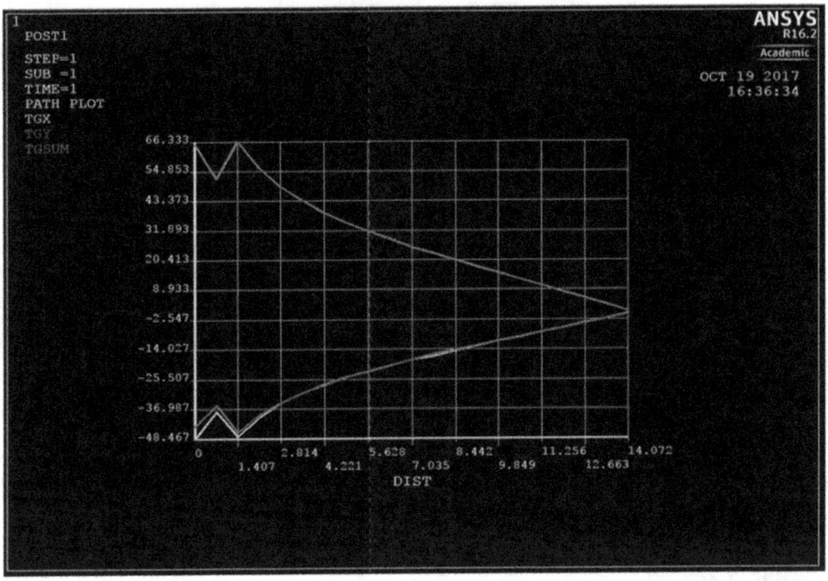

Figure 6.24. *Graph representing the variation in temperature according to the segment AB*

6.4.5. *Study of stationary thermal conduction in a reservoir intersected by a tube*

I have included a second exercise called "Steady-state thermal analysis of a pipe intersection" using the same book.

The aim of this example is to determine the temperature distribution in the intersection region between a reservoir and a tube.

The geometry studied is presented in Figure 6.25.

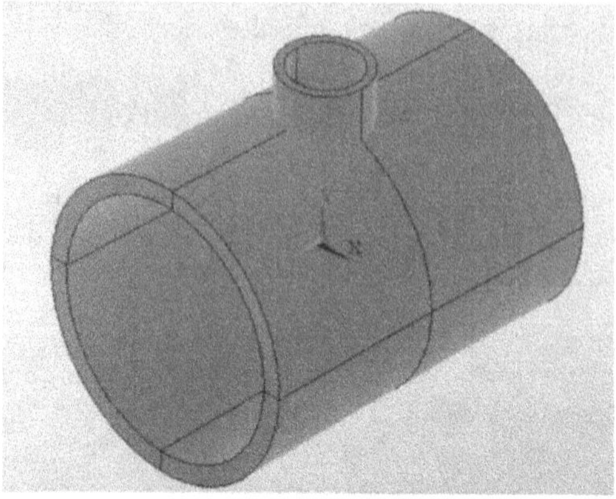

Figure 6.25. *Geometry studied: a reservoir intersected by a tube*

This is a cylindrical reservoir that is pierced radially by a small tube.

The dimensions are as follows:

– cylindrical reservoir:

- interior diameter = 26 mm,
- exterior diameter = 30 mm;

– tube:

- interior diameter = 8 mm,
- exterior diameter = 11 mm.

The temperatures imposed are as follows: a fluid temperature of 232°C, inside the reservoir, and a constant temperature of 38°C for the flux inside the tube.

Inside the reservoir, we have a convection coefficient equal to 4.92 $W/m^2°C$.

Inside the tube, depending on the temperature, the convection coefficient varies between 19.68 and 39.36 $W/m^2°C$.

The following table summarizes the values of different parameters for different known fluid temperatures (in °C).

Temperature (°C)	**21**	**93**	**149**	**204**	**260**
Coefficient convection (W/m^2°C)	41.918	39.852	34.637	27.06	21.746
Density (kg/m^3)	7,889	7,889	7,889	7,889	7,889
Conductivity (J/s m°C)	0.2505	0.267	0.2805	0.294	0.3069
Specific heat (J/kg°C)	6.898	7.143	7.265	7.448	7.631

Here, we will only model the quarter geometry resulting from the symmetries that the structure draws out.

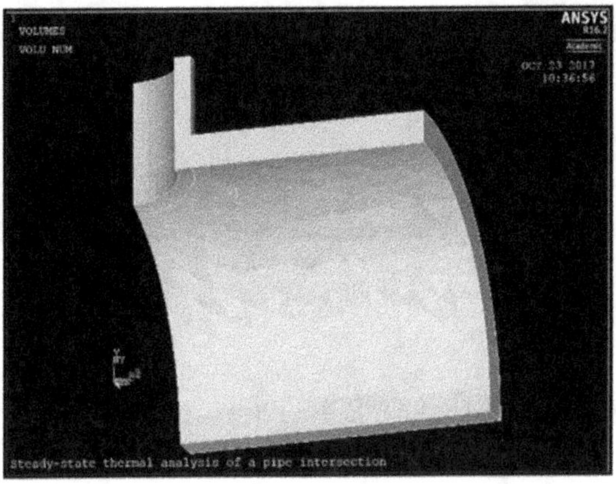

Figure 6.26. *Quarter geometry*

We chose the element type "Thermal Mass Solid, Brick 20node 90", and have included the different properties given in the above table in the software.

It should be noted that this example has enabled me to perform a study on 3D geometry and to therefore know the commands needed in order to change the view on the software.

After meshing our structure, we used the "smart mesh generator" and chose the dimension level at 1, which corresponded to a fine mesh.

The mesh geometry obtained is shown in Figure 6.27.

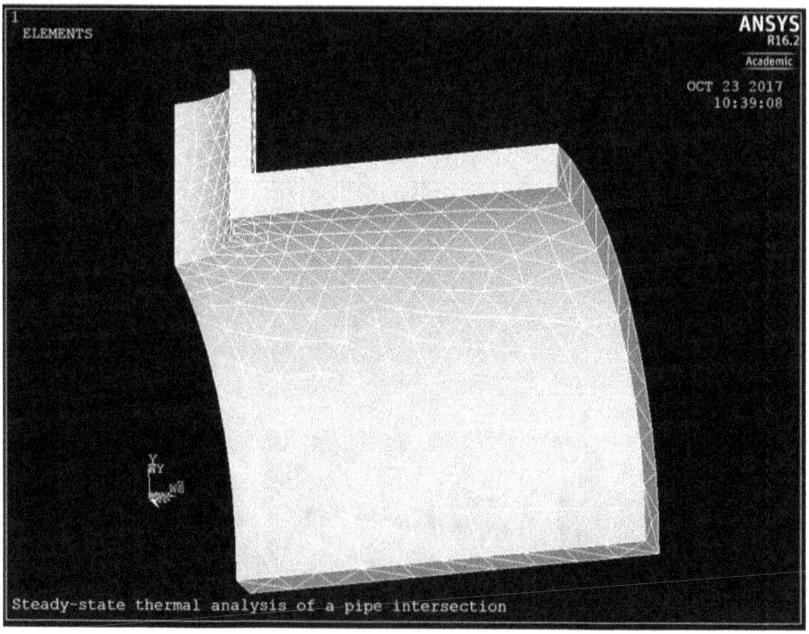

Figure 6.27. *Meshed geometry with a fine mesh (level size = 1)*

We then define our stationary type analysis using the Newton–Raphson model.

We also impose a convective load on an interior part of the small tube with:

– film coefficient = -2 W/m^2K;

– bulk temperature = 38°C.

In addition, we impose a convective load on the lower part of the reservoir with:

– film coefficient = 4.92 W/m^2K;

– bulk temperature = 232°C.

After the calculation, we obtain the results presented in Figure 6.28.

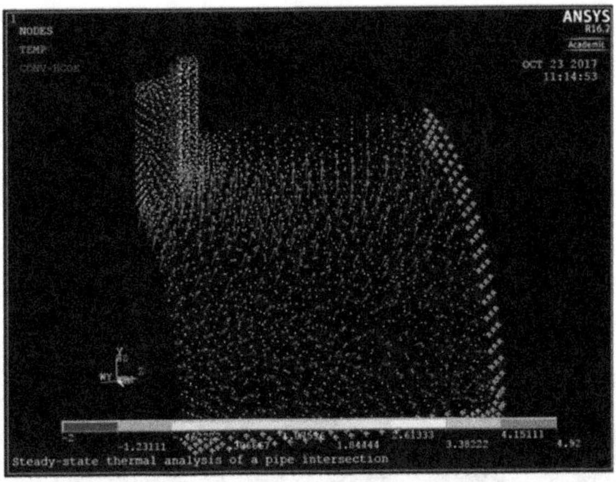

Figure 6.28. *Modeling loads on our model (with arrows)*

Figure 6.29 shows the solution of our model.

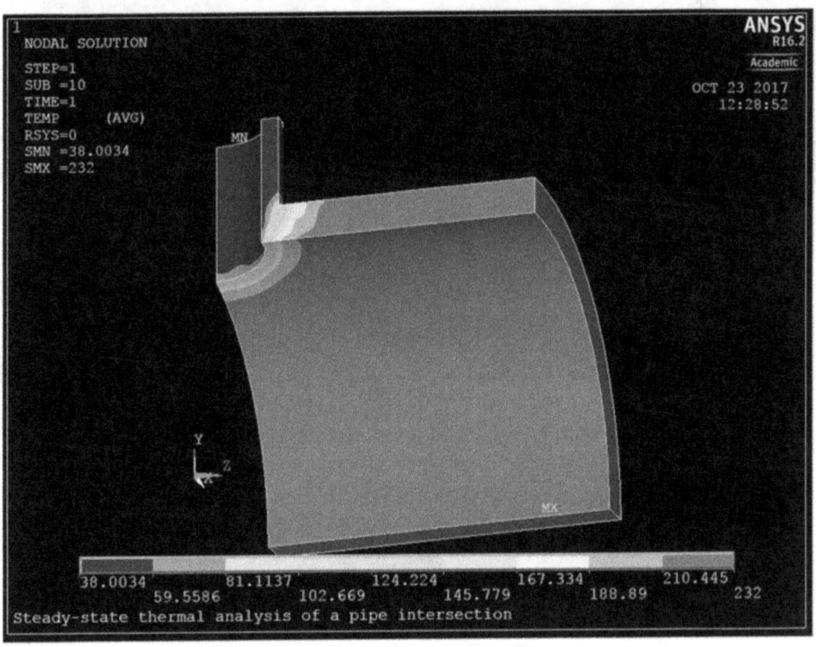

Figure 6.29. *Temperature distribution in our system*

Here, we can clearly note the temperature transfer zone between the fluids from the reservoir and the tube.

We then observe the result of the heat fluxes at the intersection between the reservoir and the tube.

To obtain a vector representation, we use the "vector plot" command with the "Raster" mode. The heat fluxes are therefore represented by vectors. The result is shown in Figure 6.30.

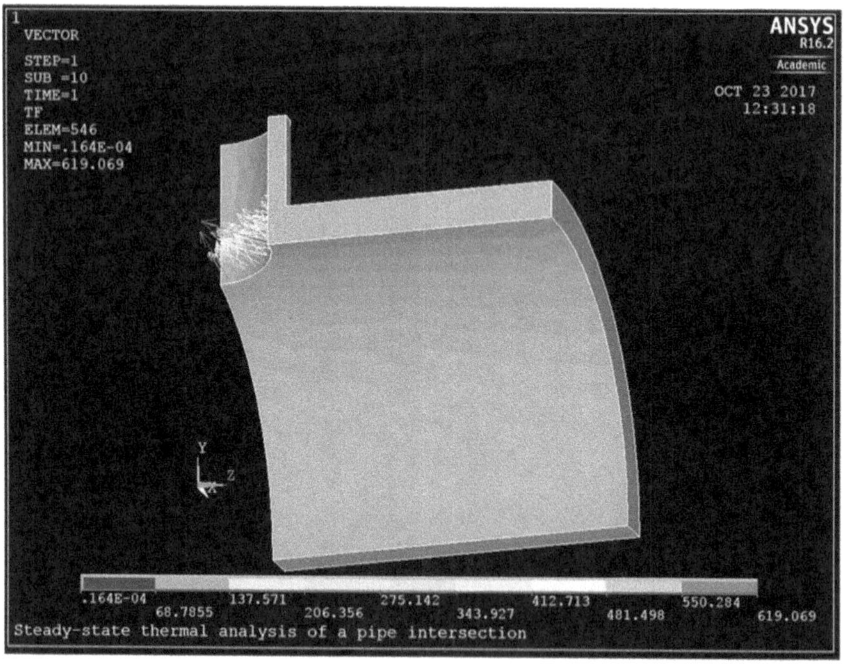

Figure 6.30. *Distribution of heat flux vectors at the intersection between the reservoir and the tube*

These vectors are directed well from the zone of high heat towards the zone with the lowest temperature.

6.4.6. *Stationary thermal conduction on a cylinder*

Problem: Observe the stationary thermal exchanges on a cylinder.

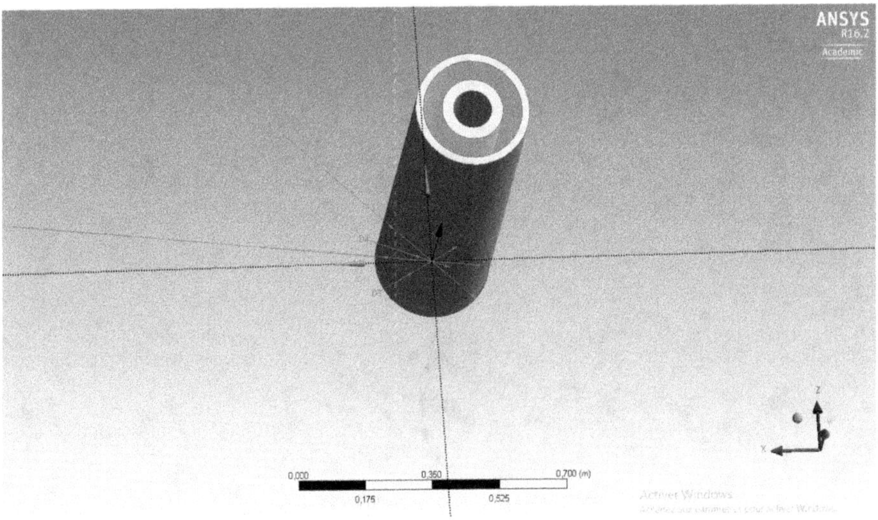

Figure 6.31. *Construction of the cylinder on a 3D model using ANSYS Workbench*

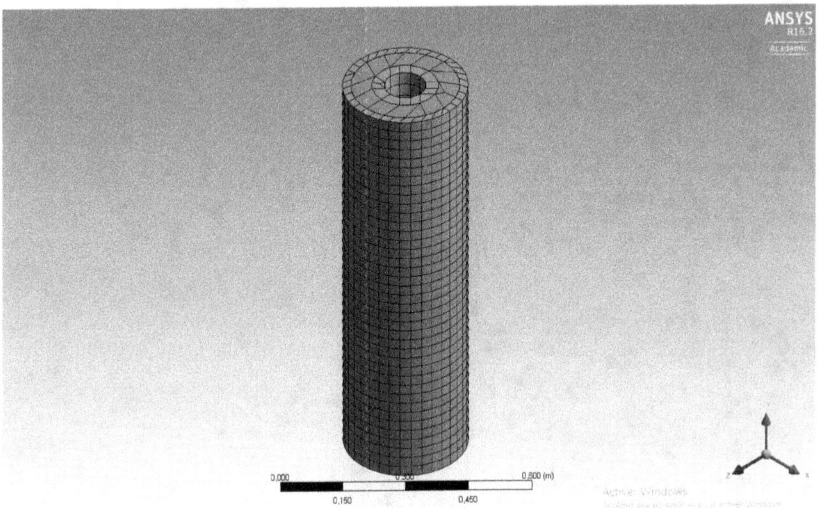

Figure 6.32. *Mesh of the cylinder*

Conditions: We applied a temperature of 375°C inside the cylinder and a temperature of 100°C on the exterior surface. Moreover, we applied a convection of 22.29 W/m^2°C.

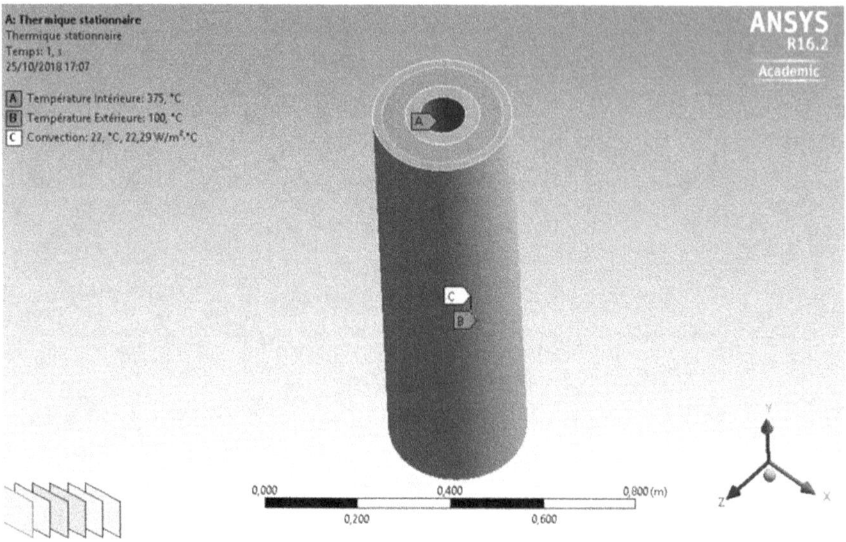

Figure 6.33. *Application of the study conditions on the cylinder*

Results of temperatures:

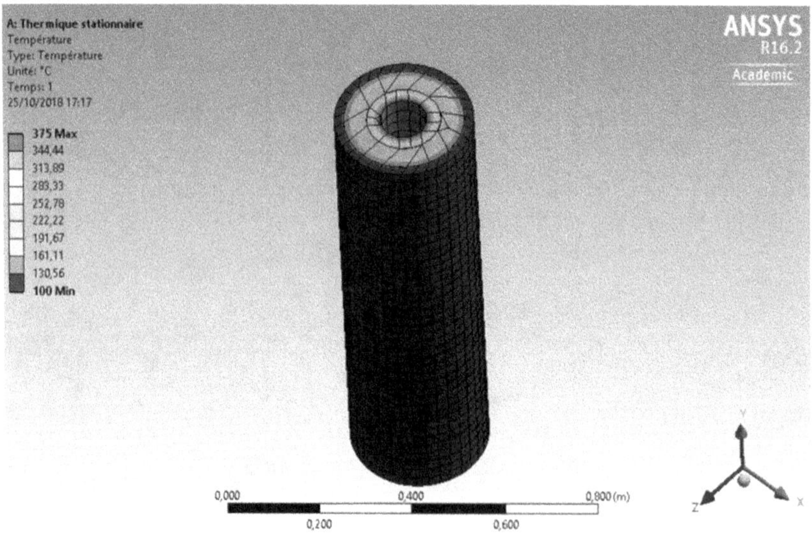

Figure 6.34. *Results of temperatures in stationary thermics on the cylinder*

Here, we can observe that the radial heat distribution occurs in a uniform manner.

Directional heat flux:

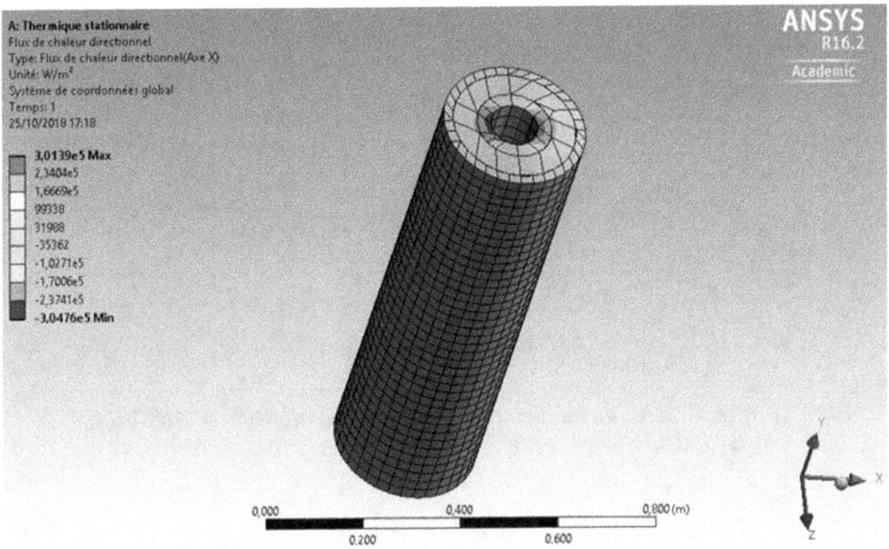

Figure 6.35. *Results of the directional heat fluxes in thermal stationary on the cylinder*

We note that the heat flux caused by this difference in temperature seems incongruous. In fact, we might think that the heat flux should only occur in a radial manner.

6.4.7. *Cooling of a puck in transitory thermal*

Problem: Observe the transitory thermal on a puck. Note how much time it takes for the puck to reach the ambient temperature of the room.

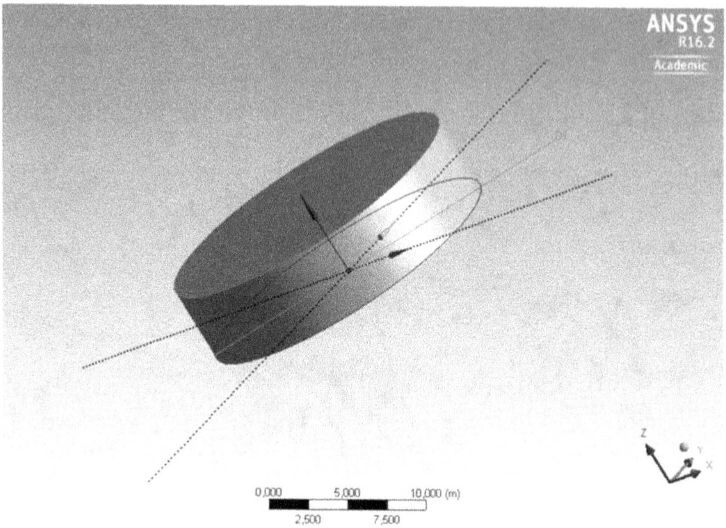

Figure 6.36. *Construction of a puck on a 3D model from ANSYS Workbench. The view of the model seen before pressure*

Conditions: We applied a convection of 10 W/m^2K between the air of the room and the puck.

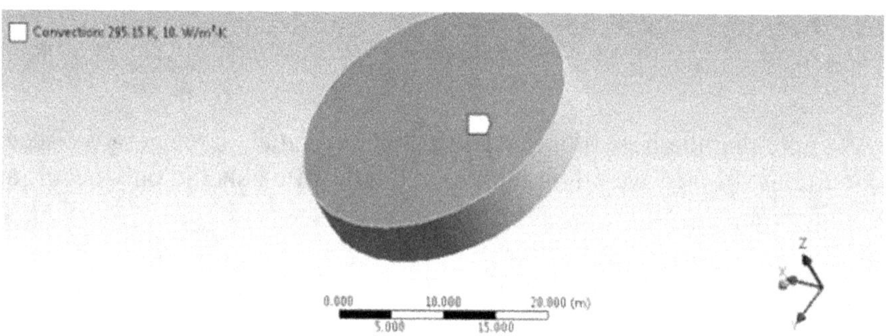

Figure 6.37. *Applying convection on the puck*

Results:

Here, are the results that we obtain for t=30 s:

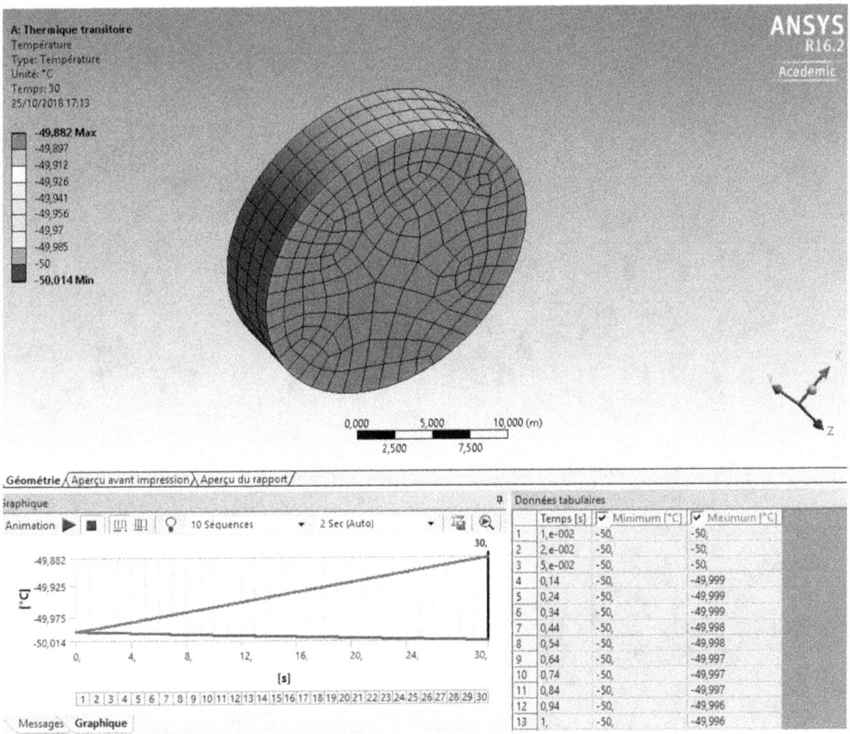

Figure 6.38. *Results in transitory thermal obtained on the puck after 30 seconds*

We note that for this very short lapse in time, the puck still did not have enough time to cool. In fact, we can observe that its maximum temperature is almost identical to its minimum temperature.

The results obtained for t=10,000 s are presented in Figure 6.39.

We can now observe that the temperature of the puck has changed all the same. However, the timespan seems to be quite long to us for this variation of only 11°C.

This may be due to the size of the puck, which is large, and the convection, which is not very strong and therefore not significant enough to observe large delta between the maximum and the minimum temperature. To observe significant cooling, it is necessary to reduce the size of our puck by running a longer simulation.

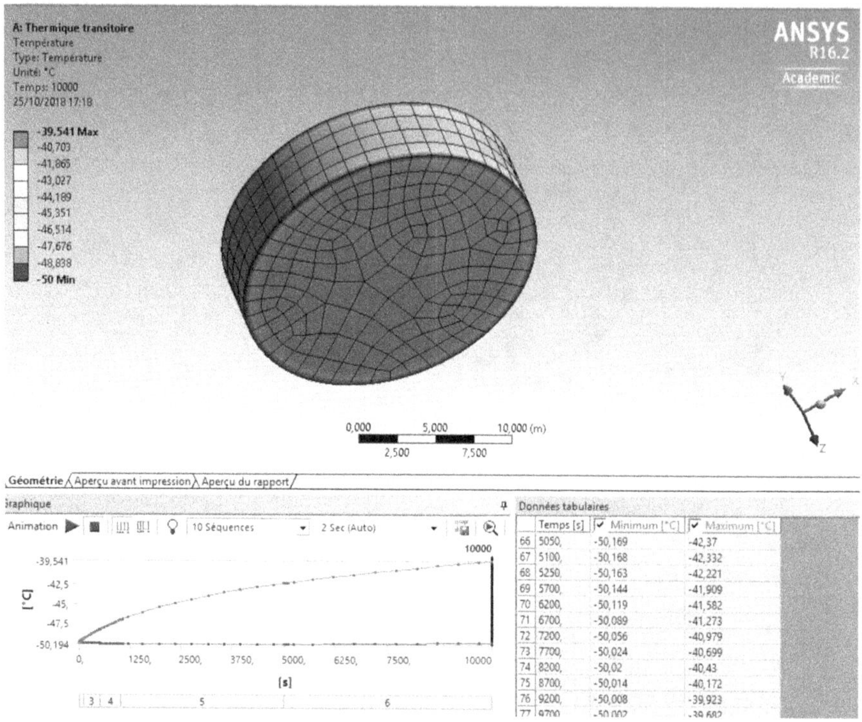

Figure 6.39. *Results in transitory thermal obtained on the puck after 10,000 seconds*

6.4.8. *Study of a heat exchanger*

We created an example in stationary thermal on a thermal exchanger without a phase exchange material.

Problem: First, observe stationary thermal exchanges on a thermal exchanger without a phase change material. Then, we will compare the differences in the results with a thermal exchanger, including a phase change material. Here, we have a forced convection with a ventilator, for example.

Conditions: We applied a significant heat flux on the lower part of the cylinders of the exchanger. We have applied convection between the exchanger and the areas of 100 W/m^{2}K. This corresponds, in fact, to a forced convection.

Results of temperatures:

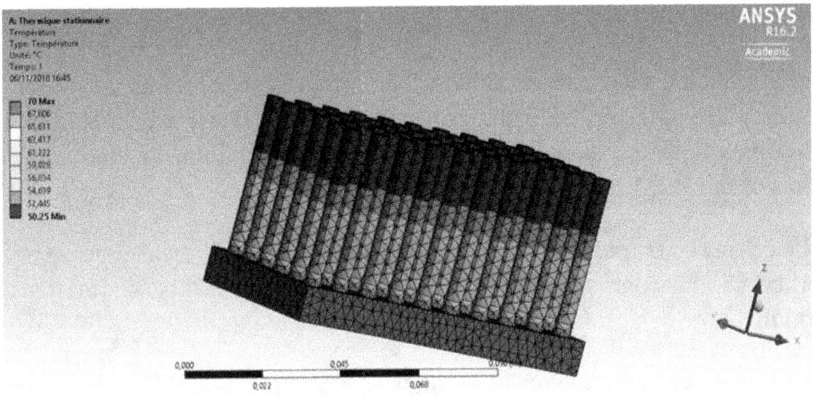

Figure 6.40. *Temperature results in thermal transitory on the thermal exchanger*

Total heat flux:

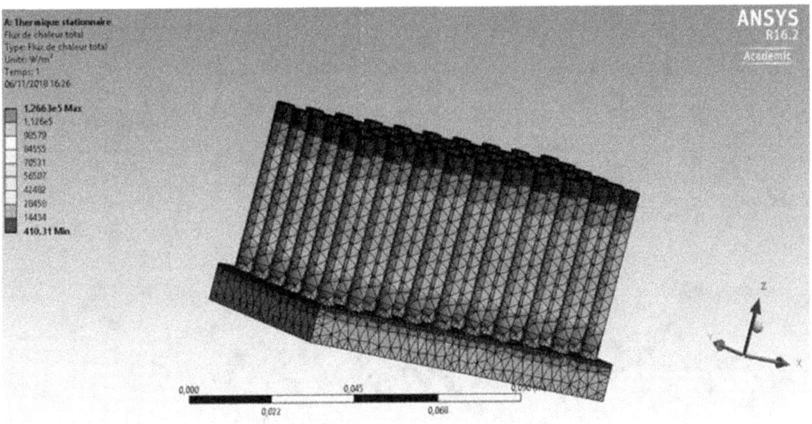

Figure 6.41. *Results of heat flux in thermal transitory on the thermal exchanger*

We can note that the highest temperature is on the lower part of the thermal exchanger, since this is where we have applied our heat flux. In addition, the total heat fluxes are more significant at the base, not far from the initially applied heat flux.

6.5. Study of a thermal exchanger on ANSYS

6.5.1. *Effectiveness of the PCM*

We modeled the analysis of the thermal exchanger using the Catia V5 software. We also modeled the part of the PCM (Phase Change Material) that comes with the exchanger using Catia, with the aim of analyzing the effects of the PCM.

Once our 3D models are constructed, we import them into ANSYS Workbench. For our simulations, we used ANSYS Workbench in transient thermal.

Our aim was to develop a simulation with and without an PCM (with different heat fluxes), in order to observe the differences in the results in transient thermal.

6.5.2. *Parameterizing the analysis*

To our solids, we applied a heat flux in order to simulate the heat extricated by an electronic chip, a convection and a radiation.

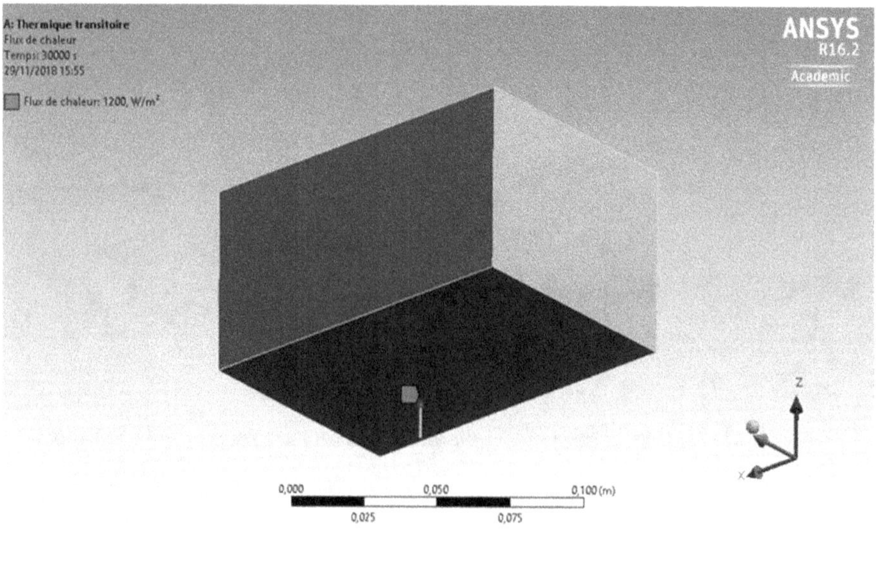

Figure 6.42. *Parameterizing of a heat flux to simulate the heat extricated by an electronic chip*

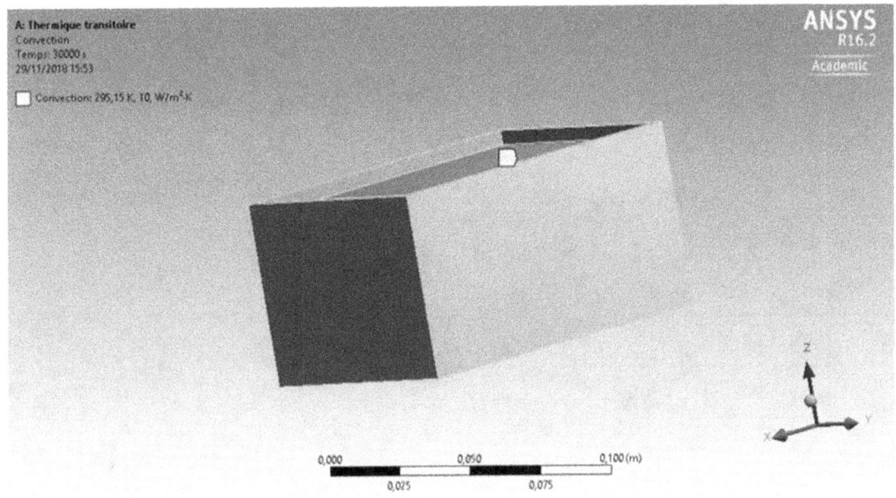

Figure 6.43. *Parameterizing a convection*

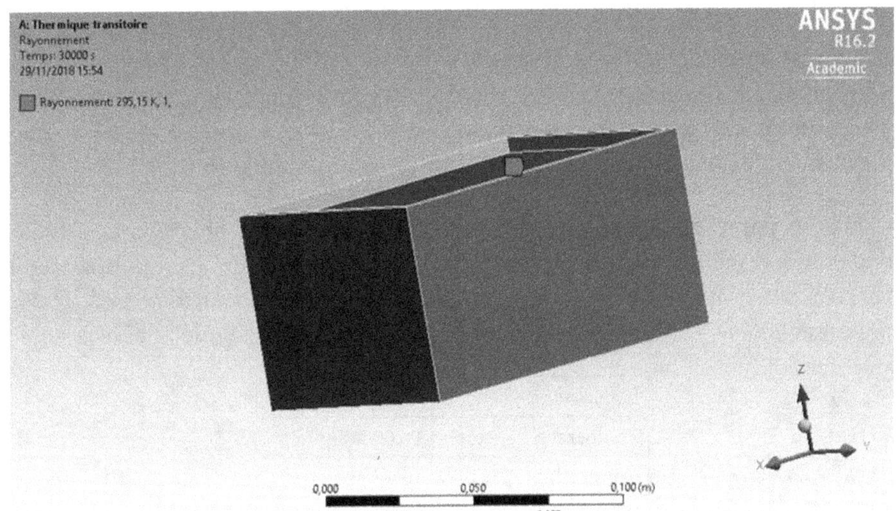

Figure 6.44. *Parameterizing the radiation*

We also created a mesh on our solids before starting the analyses. Our element size was $3*10^{-3}$ m, with a maximum stopping length of $5*10^{-3}$ m, all with a medium smoothing.

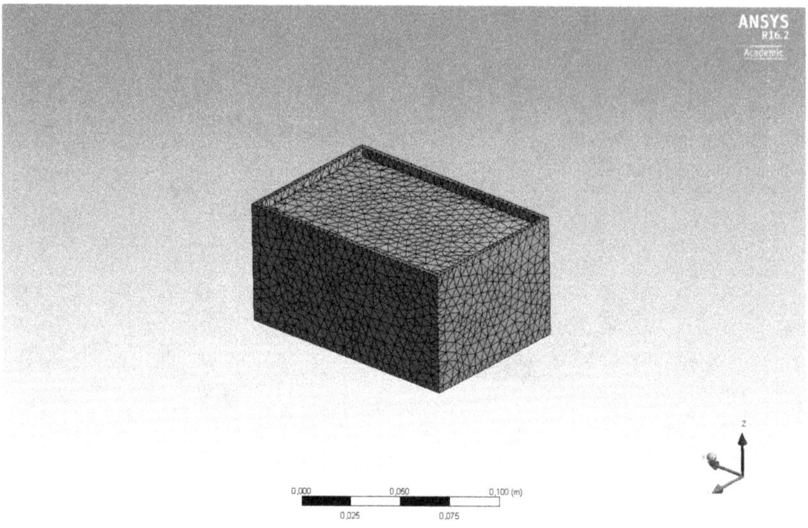

Figure 6.45. *Development of a mesh with ANSYS Workbench*

We parameterized our simulations with an initial time increment of 0.1 s, a minimal time increment of 0.1 s and a maximum time increment of 100 s. These parameters are very significant, so they have a precise curve during phase changes (significant, above all, for the PCM simulation).

These parameters will, in fact, allow us to have a convergence of our calculations without taking up too much computing time. This is important for us, since we may have a significant number of calculations. These parameters (see Table 6.1) will be used for the following three analyses.

Parameters	Values
Temperature fluxes	1,200 W/m^2
Convection	10 W/m^{2}K
Radiation	Emissivity=1
Meshing	3 E-3
Analysis time	30,000 seconds

Table 6.1. *Parameters and associated values*

6.5.3. *Thermal exchanger without an PCM*

The first simulation was performed without an MCP whose configuration is presented in Figure 6.46, showing the heat flux of the thermal exchanger (temperature vs. time: the temperature increases from the ambient temperature (297 K) up to the permanent working temperature (325 K)). Initially, the temperature increases significantly and then begins to decrease with time. This result is not surprising, since at the surface temperature, on which the heat flux is applied, the difference between the temperature and the ambient temperature (Ts-T∞) increases. This means that the heat rejected by convection and radiation increases with time and is close to the heat generated by the surface of the base of the thermal exchanger, in order to reach the state of equilibrium at 1,200 W/m^2 after t=9,000 s (which is 150 minutes), as shown in our graph. Eventually, the heat generated becomes equal to the heat rejected.

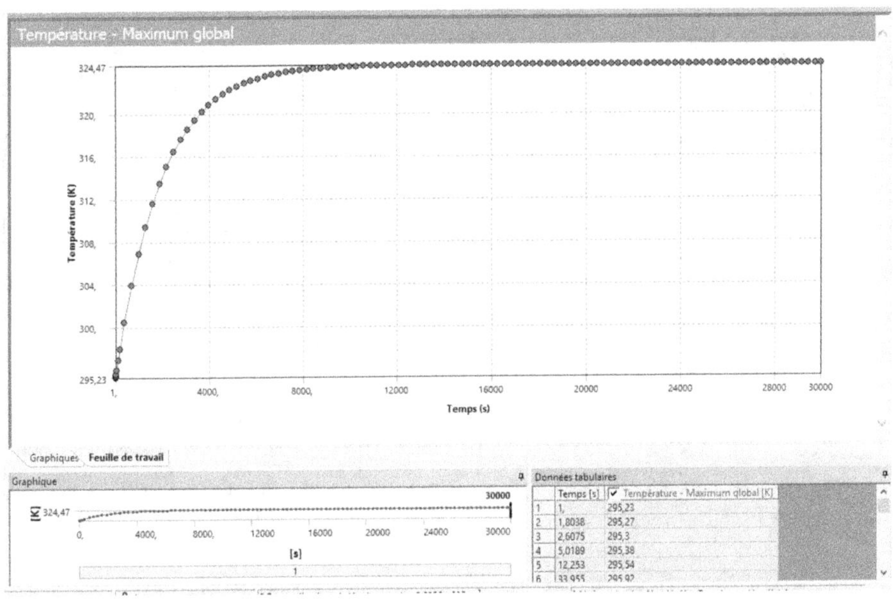

Figure 6.46. *Temperature versus time for a simulation without a PCM*

6.5.4. *Thermal exchanger with hydrated salt*

The choice of a PCM is crucial for this type of study; the PCM must have a solidus temperature above the ambient temperature and a liquidus temperature below that of the stationary state of the thermal exchanger. In addition, the density of the PCM to be selected is important since we are limited by the volume used in the exchanger. Therefore, a PCM with a high density leads to a greater solid-to-liquid mass transfer and vice versa. Here, the PCM used in this simulation is a hydrated salt. Other parameters such as latent heat, specific heat, or even thermal conductivity, can be taken into account.

To use a PCM with ANSYS Workbench in transitory thermal, we created a material in the library. Then, to characterize the material, we researched the density, the isotropic thermal conductivity and the specific heat. However, we could not provide direct information on the latent fusion, so we had to calculate the enthalpy depending on the temperature at different strategic points, and enter this in ANSYS in table form.

The thermo-physical properties of the hydrated salt used for our analysis are presented in Table 6.2.

Density (kg/m^3)	1,500
Viscosity (kg/m.s)	0.00184
Thermal conductivity (W/m.K)	0.6
Specific thermal capacity (kJ/kg.K)	2
Solidus temperature (°C)	27
Liquidus temperature (°C)	32
Latent heat fusion (kJ/kg)	200

Table 6.2. *Thermo-physical properties of hydrated salt*

Now, instead of air filling the spaces between the fins, the PCM is modeled using ANSYS and placed between the fins across the whole length. Throughout the PCM phase change, the volume is considered constant, and the melted PCM does not slide outside the thermal dissipater.

To study the effect of the MCP on the transitory variation of thermal exchanger temperature (with and without an PCM), we have presented two different graphs.

Our graphs clearly show that, initially, the temperature of the thermal dissipator with and without the PCM increases in the same way, to reach the fusion temperature of the MCP. At this stage, the salt maintains its temperature, while the temperature of the thermal dissipator (without the PCM) is maintained by increasing it until it reaches the steady-state temperature (324 K). The time needed for the thermal dissipator with the PCM to reach the equilibrium temperature varies from 9,000 to 30,000 s. Thus, during the fusion process, the PCM delays the maximum temperature of the dissipator by absorbing the energy produced by the electronic chip at work.

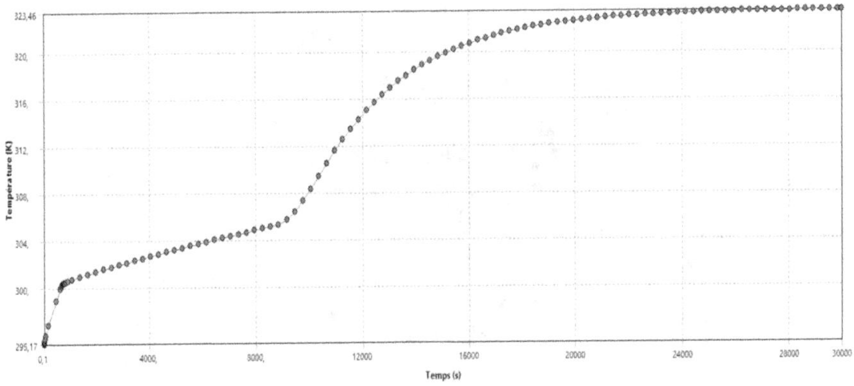

Figure 6.47. *Temperature results for the hydrated PCM salt with a heat flux of 1,200 W/m*^{2}

In fact, we can note in our results that the lower part on which the heat flux is applied is the hottest part of the system, and the further we go from this zone, the more the temperature decreases.

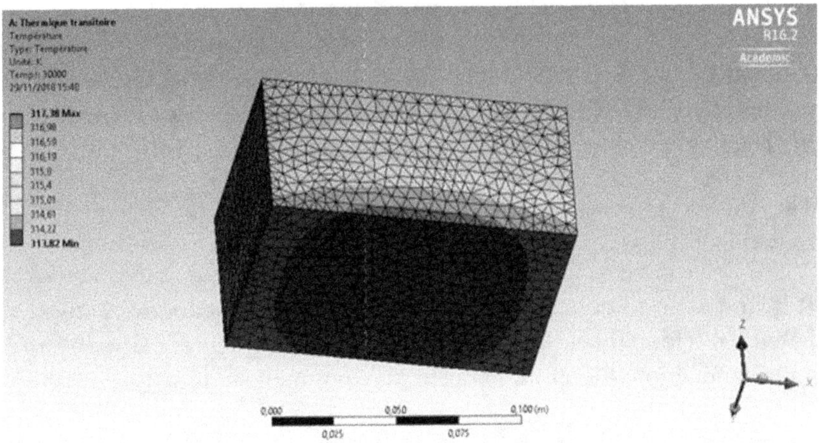

Figure 6.48. *Representation of heat on the thermal exchanger (seen from below)*

The PCM is the coldest area of the system, since it uses the heat received to change the state.

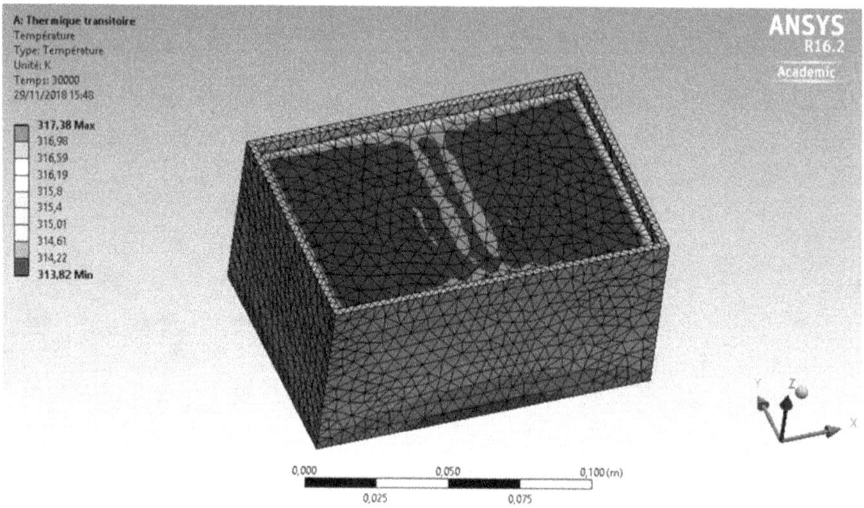

Figure 6.49. *Representation of heat on the thermal exchanger seen from above)*

6.5.5. *Thermal exchanger with paraffin*

As we mentioned above, the choice of an adapted PCM is primordial; hence, we found it pertinent to compare our results with a PCM. The results were very close where specific heat and latent heat fusion were concerned, but had a different density and thermal conductivity. In fact, the density is an important factor to take into account as it directly influences the mass of the PCM. The phase change material that we have chosen is paraffin.

The thermo-physical properties of paraffin used in our analysis are presented in Table 6.3.

It is important to note that paraffin has a thermal conductivity lower than hydrated salt (four times lower), so this PCM can theoretically absorb less heat via conduction, therefore paraffin may not melt so quickly.

These hypotheses have been verified by our simulation using ANSYS.

Density (kg/m^3)	820
Viscosity (kg/m.s)	0.00068
Thermal conductivity (W/m.K)	0.15
Specific thermal capacity (kJ/kg.K)	2.2
Solidus temperature (°C)	30
Liquidus temperature (°C)	40
Latent heat fusion (kJ/kg)	190

Table 6.3. *Thermo-physical properties of paraffin*

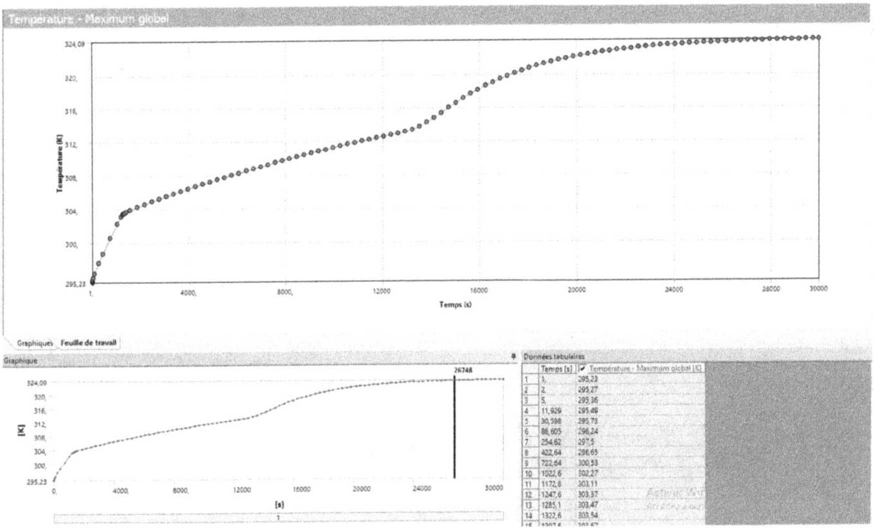

Figure 6.50. *Temperature results for paraffin as a MCP with a heat flux of 1,200 W/m^2*

Here, the temperature reached 324 K after 26,000 s with paraffin compared to 30,000 s with hydrated salt. In addition, we note that the maximum temperature reached with hydrated salt was 323 K compared to 324 K with paraffin, which may be negligible.

6.5.6. *Influence of heat flux*

We found it pertinent to perform the simulation using an MCP with applied heat fluxes, in order to observe the effect of this factor on the system. We therefore rerun two other simulations with 3,000 W/m^2 and 6,000 m^2.

The results obtained are summarized in Figure 6.51.

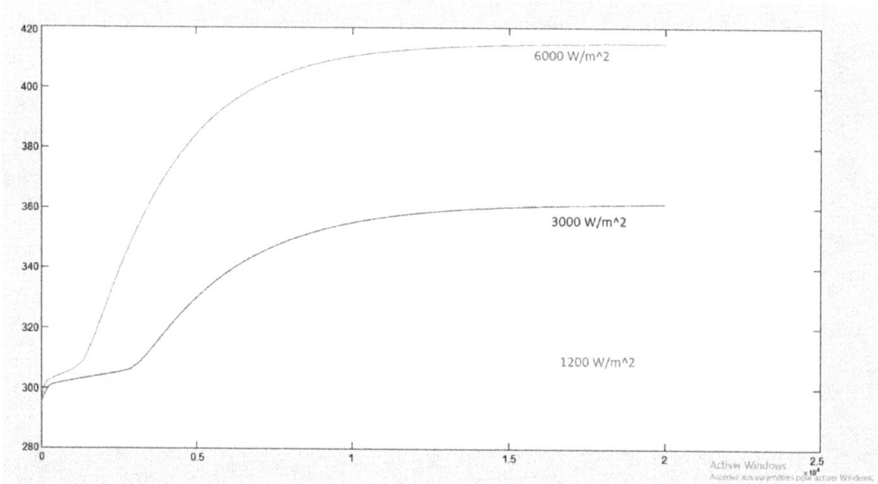

Figure 6.51. *Graphical representation of the different heat fluxes applied for hydrated salt*

By comparing our three different heat fluxes, we can note that the stationary temperature is reached more rapidly when the heat flux is significant. The MCP absorbs more heat and melts more quickly. Its effect is very significant for a heat flux of 1,200 W/m^2, since we note a more significant plateau before the temperature increases again. However, for greater heat fluxes such as 3,000 W/m^2 and 6,000 W/m^2, the plateau decreases; therefore, the effect of the MCP needs to be nuanced further.

As with hydrated salt, we carried out the analysis with paraffin at different values of heat flux. The measured values were the same as for hydrated salt, which were 1,200 W/m^2, 3,000 W/m^2 and 6,000 W/m^2.

The curves based on these values are presented in Figure 6.52.

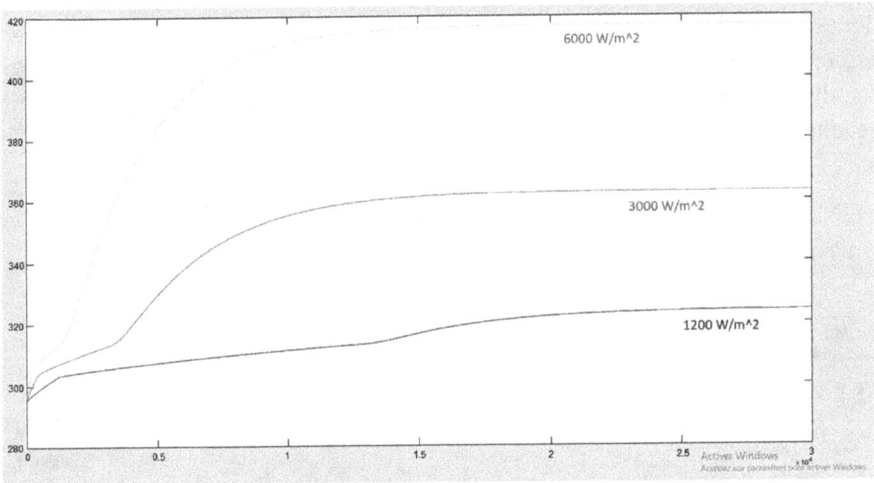

Figure 6.52. *Graphical representation of the different heat fluxes applied for paraffin*

We can note that the results obtained seem quite similar to those with hydrated salt. This means that the stationary temperature is reached more quickly when the heat flux is significant.

The results clearly reveal that hydrated salt has a better performance in delaying the temperature peak than paraffin.

6.5.7. *Comparing PCM*

The aim of this section is to carry out a study on paraffin, hydrated salt and the thermal exchanger alone, in order to find whether there is any influence on the three configurations throughout the time cycle of 30,000 s.

We can note that the effectiveness of MCPs lies in delaying the final temperature, rather than the thermal exchanger alone. However, these do not react in the same way. In the case of hydrated salt, we can note that it stabilizes the temperature much slower when compared with paraffin, which is due to its lower fusion temperature. However, the duration at which the temperature is more or less stabilized is higher than that for paraffin, which is due to its broad fusion temperature range, although its latent heat is slightly lower. We therefore note that from 12,000 seconds, the temperature curve of hydrated salt rises above that of paraffin. In fact, from this result, we can deduce that for a temperature lower than 312 K, hydrated salt is more effective and the reverse is true for paraffin. Nevertheless, it is important to recall that paraffin reaches 324 K after 26,000 s, given that hydrated salt reaches 324 K a little later (about 30,000 s).

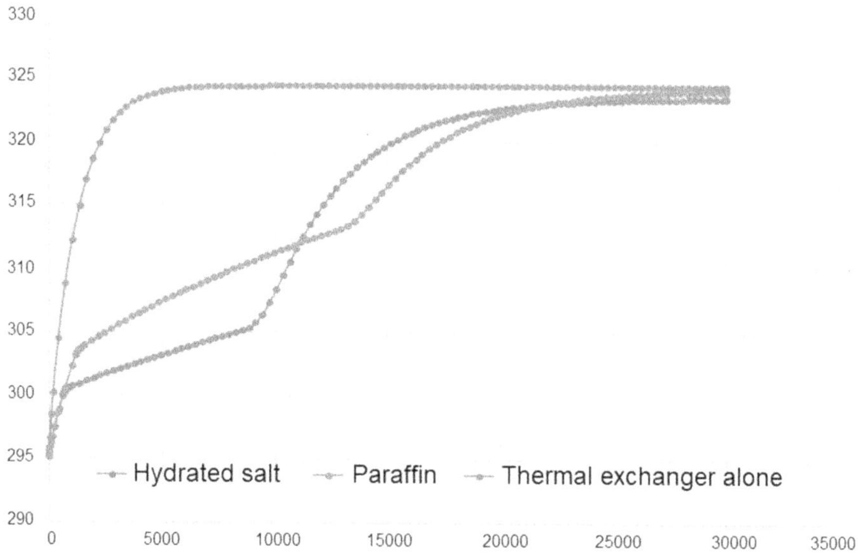

Figure 6.53. *Comparison of the three models*

6.6. Conclusion

The phase change material has become more widespread in the field of building insulation; however, it is still only used occasionally in the field of electronic technologies. We have seen that the phase change material can provide a response that is not only effective for thermomechanical problems encountered in the electronic domain, which will prevent electronic systems from degrading. It also consumes less energy and is therefore more ethical. In a world increasingly damaged by human activity, we believe that the technologies of the future should play a role in a responsible approach to industrial ecology.

However, for these phase change materials to be used more often in electronics, there is some progression to be made, especially in improving the thermal conductivity of these materials, since it is vital to find a thermal conductivity adapted to the system. In fact, the MCP chosen should meet specifications and constraints imposed by the system. Moreover, one of the drawbacks of MCPs is still crystallization kinetics, which is relatively weak.

As for us, this project allowed us to get as close as possible to a problem that the industry is encountering today and reflect on how to provide a solution. We were led to use different software such as ANSYS and even Matlab, which enabled us to deepen our understanding using a practical case. This was a valuable experience that will help us in our future work as engineers.

Appendix

$G_{0-\lambda T}$ Function Table

Remember that $G_{0-\lambda T}$ is the fraction of radiation from a black body at temperature T between wavelengths 0 and λ, which is:

$$G_{0-\lambda T} = \frac{1}{\sigma T^4} \int_0^{\lambda} \frac{C_1 \, d\dfrac{u}{T}}{\left(\dfrac{u}{T}\right)^5 \left[\exp\left(\dfrac{C_2}{u}\right) - 1\right]} = \frac{C_1}{\sigma} \int_{u=0}^{u=\lambda T} \frac{du}{u^5 \left[\exp\left(\dfrac{C_2}{u}\right) - 1\right]} \qquad [A.1]$$

It is easy to use this table.

We should recall that λT should be calculated by expressing wavelengths λ in micrometers (μm), and temperatures T in Kelvin (K).

EXAMPLE.– What is the function $G_{0-\lambda T}$ for $\lambda = 500\ nm$ and $T = 4440\ K$?

$$\lambda = 500\ nm = 0.500\ \mu m,$$

$$\text{so } \lambda T = 0.5 * 4440 = 2220\ \mu m.K \qquad [A.2]$$

To determine G_{0-2220}, we find the line $\lambda T = 2200$ in the table and take the value in the column marked 20.

We then find:

$$G_{0-2220} = 0.1045 \quad [A.3]$$

λT	0	20	40	60	80
1700	0.0285	0.0305	0.0326	0.0347	0.0370
1800	0.0393	0.0471	0.0442	0.0467	0.0494
1900	0.0521	0.0549	0.0577	0.0606	0.0636
2000	0.0667	0.0698	0.0730	0.0763	0.0796
2100	0.0830	0.0865	0.0900	0.0936	0.0972
2200	0.1009	0.1045	0.1084	0.1122	0.1191
2300	0.1200	0.1240	0.1280	0.1320	0.1361
2400	0.1402	0.1444	0.1486	0.1528	0.1571
2500	0.1613	0.1656	0.1700	0.1743	0.1787
2600	0.1831	0.1875	0.1920	0.1964	0.2009
1700	0.0285	0.0305	0.0326	0.0347	0.0370
500	0.0000	0.0000	0.0000	0.0000	0.0000
600	0.0000	0.0000	0.0000	0.0000	0.0000
700	0.0000	0.0000	0.0000	0.0000	0.0000
800	0.0000	0.0000	0.0000	0.0000	0.0000
900	0.0001	0.0001	0.0001	0.0001	0.0002
1000	0.0003	0.0004	0.0004	0.0005	0.0007
1100	0.0009	0.0010	0.0013	0.0015	0.0018
1200	0.0021	0.0024	0.0028	0.0033	0.0037
1300	0.0043	0.0049	0.0055	0.0062	0.0069
1400	0.0078	0.0086	0.0096	0.0106	0.0117
1500	0.0128	0.0140	0.0153	0.0167	0.0182
1600	0.0197	0.0213	0.0230	0.0247	0.0266
1700	0.0285	0.0305	0.0326	0.0347	0.0370
1800	0.0393	0.0471	0.0442	0.0467	0.0494
1900	0.0521	0.0549	0.0577	0.0606	0.0636
2000	0.0667	0.0698	0.0730	0.0763	0.0796
2100	0.0830	0.0865	0.0900	0.0936	0.0972
2200	0.1009	0.1045	0.1084	0.1122	0.1191
2300	0.1200	0.1240	0.1280	0.1320	0.1361

λT	**0**	**20**	**40**	**60**	**80**
2400	0.1402	0.1444	0.1486	0.1528	0.1571
2500	0.1613	0.1656	0.1700	0.1743	0.1787
2600	0.1831	0.1875	0.1920	0.1964	0.2009
2700	0.2053	0.2098	0.2143	0.2188	0.2234
2800	0.2279	0.2324	0.2369	0.2415	0.2460
2900	0.2506	0.2551	0.2596	0.2642	0.2687
3000	0.2732	0.2778	0.2823	0.2868	0.2913
3100	0.2958	0.3003	0.3047	0.3092	0.3137
3200	0.3181	0.3225	0.3269	0.3313	0.3357
3300	0.3401	0.3445	0.3488	0.3531	0.3574
3400	0.3617	0.3660	0.3703	0.3745	0.3787
3500	0.3829	0.3871	0.3912	0.3954	0.3995
3600	0.4036	0.4077	0.4117	0.4153	0.4198
3700	0.4238	0.4277	0.4317	0.4356	0.4395
3800	0.4434	0.4472	0.4511	0.4549	0.4585
3900	0.4624	0.4661	0.4699	0.4736	0.4772
4000	0.4809	0.4845	0.4881	0.4917	0.4952
4100	0.4987	0.5022	0.5057	0.5092	0.5126
4200	0.5160	0.5194	0.5227	0.5261	0.5294
4300	0.5327	0.5359	0.5392	0.5424	0.5456
4400	0.5488	0.5519	0.5551	0.5582	0.5612
4500	0.5643	0.5673	0.5703	0.5733	0.5763
4600	0.5793	0.5822	0.5851	0.5880	0.5908
4700	0.5937	0.5965	0.599.	0.6020	0.6048
4800	0.6075	0.6102	0.6129	0.6156	0.6182
4900	0.6209	0.6235	0.6261	0.6286	0.6312
5000	0.6337	.06362	0.6387	0.6412	0.6436
5100	0.6451	0.6485	0.6509	0.6532	0.6556
5200	0.6579	0.6603	0.6625	0.6648	0.6671
5300	0.6693	0.6716	0.6738	0.6760	0.6782
5400	0.6803	0.6825	0.6845	0.6867	0.6888
5500	0.6909	0.6929	0.6950	0.6970	0.6990
5600	0.7010	0.7030	0.7049	0.7069	0.7088

λT	**0**	**20**	**40**	**60**	**80**
5700	0.7107	0.7126	0.7145	0.7164	0.7183
5800	0.7201	0.7219	0.7238	0.7256	0.7273
5900	0.7291	0.7309	0.7326	0.7343	0.7361
6000	0.7378	0.7395	0.7411	0.7428	0.7444
6100	0.7461	0.7477	0.7493	0.7509	0.7525
6200	0.7541	0.7556	0.7572	0.7587	0.7603
6300	0.7618	0.7633	0.7648	0.7662	0.7677
6400	0.7692	0.7706	0.7721	0.7735	0.7749
6500	0.7763	0.7777	0.7791	0.7804	0.7818
6600	0.7831	0.7845	0.7858	0.7871	0.7884
6700	0.7897	0.7910	0.7923	0.7936	0.7948
6800	0.7961	0.7973	0.7985	0.7998	0.8010
6900	0.8022	0.8034	0.8045	0.8057	0.8069
7000	0.8080	0.8092	0.8103	0.8115	0.8126
7100	0.8137	0.8148	0.8159	0.8170	0.8181
7200	0.8191	0.8202	0.8213	0.8223	0.8234
7300	0.8244	0.8254	0.8264	0.8275	0.8285
7400	0.8295	0.8304	0.8314	0.8324	0.8334
7500	0.8343	0.8353	0.8362	0.8372	0.8381
7600	0.8390	0.8399	0.8409	0.8418	0.8427
7700	0.8436	0.8444	0.8453	0.8462	0.8471
7800	0.8479	08488	0.8496	0.8505	0.8513
7900	0.8521	0.8530	0.8538	0.8546	0.8554
8000	0.8562	0.8570	0.8578	0.8586	0.8594
8100	0.8601	0.8609	0.8617	0.8624	0.8632
8200	0.8639	0.8647	0.8654	0.8661	0.8669
8300	0.8676	0.8683	0.8690	0.8697	0.8704
8400	0.8711	0.8718	0.8725	0.8732	0.8738
8500	0.8745	0.8752	0.8759	0.8765	0.8772
8600	0.8778	0.8785	0.879 1	0.8797	0.8804
8700	0.8810	0.8816	0.8822	0.8829	0.8835
8800	0.8841	0.8847	0.8853	0.8859	0.8865
9000	0.8889		0.8911		0.8922

λT	**0**	**20**	**40**	**60**	**80**
9100		0.8933		0.8944	
9200	0.8955		0.8965		0.8976
9300		0.8986		0.8996	
9400	0.9006		0.9016		0.9026
9500		0.9035		0.9045	
9600	0.9054		0.9063		0.9072
9700		0.9081		0.9090	
9800	0.9099		0.9108		0.9116
9900		0.9125		0.9133	
10000	0.9142				
10200	0.9181				
10400	0.9219				
10600	0.9254				
10800	0.9287				
11000	0.9318				
14000	0.9629				
16000	0.9737				
18000	0.808				
20000	0.9855				
40000	0.9978				
60000	0.9993				
80000	0.9997				
100000	0.9998				

References

[BYR 66] BYRON BIRD R., STEWART W.E., LIGHTFOOT E.N., *Transport Phenomena*, John Wiley and Sons, New York 1966.

[COS 97] COSAR P., *Aide mémoire du thermicien*, Amicale des anciens élèves de l'Ecole de thermique, Paris, 1997.

[ELH 13] EL HAMI A., RADI B., *Incertitude et optimisation et fiabilité des structures*, Hermes-Lavoisier, Paris, 2013.

[GHA 91] GHANEM R.G., SPANOS P.D., *Stochastic Finite Elements: A Spectral Approach, Revised Edition*, Springer Verlag, Berlin, 1991.

[GOS 93] GOSSE J., *Guide technique de thermique*, Dunod, Paris, 1993.

[HOL 86] HOLMAN J.P., *Heat Transfer*, McGraw Hill, New York 1986.

[JAN 16] JANNOT Y., MOYNE C., *Transferts thermiques*, Edilivres, Paris, 2016.

[KNU 58] KNUDSEN J.G., KATZ D.L., *Fluid Dynamics and Heat Transfer*, McGraw Hill, New York, 1958.

[LEO 79] LEONTIEV A., *Théorie des échanges de chaleur et de masse*, Editions MIR, Moscow, 1979.

[PER 97] PEREZ J.P., *Thermodynamique, fondements et applications*, Masson, Paris, 1997.

[RAD 16] RADI B., EL HAMI A., *Material Forming Processes: Simulation, Drawing, Hydroforming and Additive Manufacturing*, ISTE Ltd, London and John Wiley & Sons, New York, 2016.

[SAC 15] SACCADURA J.F., *Transferts thermiques, initiation et approfondissement*, Lavoisier Editeur, Paris, 2015.

[VAN 76] VAN WYLEN G., SONNTAG R.E., *Fundamentals of Classical Thermodynamics, SI Version*, John Wiley and Sons, New York 1976.

[WAR 98] WARREN M., ROSENHOW J.P., HARTNETT Y.I. *et al.*, *Handbook of Heat Transfer*, McGraw Hill, 1998.

Index

A, B, C

E, F, G

I, L, O, P

R, S, T, W

Other titles from

iSTE

in

Mechanical Engineering and Solid Mechanics

2021

DAHOO Pierre Richard, POUGNET Philippe, EL HAMI Abdelkhalak
Applications and Metrology at Nanometer Scale 1: Smart Materials, Electromagnetic Waves and Uncertainties
(Reliability of Multiphysical Systems Set – Volume 9)

2020

LEDOUX Michel, EL HAMI ABDELKHALAK
Heat Transfer 1: Conduction

SALENÇON Jean
Elastoplastic Modeling

2019

BAYLE Franck
Reliability of Maintained Systems Subjected to Wear Failure Mechanisms: Theory and Applications
(Reliability of Multiphysical Systems Set – Volume 8)

BEN KAHLA Rabeb, BARKAOUI Abdelwahed, MERZOUKI Tarek
Finite Element Method and Medical Imaging Techniques in Bone Biomechanics
(Mathematical and Mechanical Engineering Set – Volume 8)

IONESCU Ioan R., QUEYREAU Sylvain, PICU Catalin R., SALMAN Oguz Umut
Mechanics and Physics of Solids at Micro- and Nano-Scales

LE VAN Anh, BOUZIDI Rabah
Lagrangian Mechanics: An Advanced Analytical Approach

MICHELITSCH Thomas, PÉREZ RIASCOS Alejandro, COLLET Bernard, NOWAKOWSKI Andrzej, NICOLLEAU Franck
Fractional Dynamics on Networks and Lattices

SALENÇON Jean
Viscoelastic Modeling for Structural Analysis

VÉNIZÉLOS Georges, EL HAMI Abdelkhalak
Movement Equations 5: Dynamics of a Set of Solids
(Non-deformable Solid Mechanics Set – Volume 5)

2018

BOREL Michel, VÉNIZÉLOS Georges
Movement Equations 4: Equilibriums and Small Movements
(Non-deformable Solid Mechanics Set – Volume 4)

FROSSARD Etienne
Granular Geomaterials Dissipative Mechanics: Theory and Applications in Civil Engineering

RADI Bouchaib, EL HAMI Abdelkhalak
Advanced Numerical Methods with Matlab® 1: Function Approximation and System Resolution
(Mathematical and Mechanical Engineering SET – Volume 6)
Advanced Numerical Methods with Matlab® 2: Resolution of Nonlinear, Differential and Partial Differential Equations
(Mathematical and Mechanical Engineering SET – Volume 7)

SALENÇON Jean
Virtual Work Approach to Mechanical Modeling

2017

BOREL Michel, VÉNIZÉLOS Georges
Movement Equations 2: Mathematical and Methodological Supplements (Non-deformable Solid Mechanics Set – Volume 2)
Movement Equations 3: Dynamics and Fundamental Principle (Non-deformable Solid Mechanics Set – Volume 3)

BOUVET Christophe
Mechanics of Aeronautical Solids, Materials and Structures
Mechanics of Aeronautical Composite Materials

BRANCHERIE Delphine, FEISSEL Pierre, BOUVIER Salima, IBRAHIMBEGOVIC Adnan
From Microstructure Investigations to Multiscale Modeling: Bridging the Gap

CHEBEL-MORELLO Brigitte, NICOD Jean-Marc, VARNIER Christophe
From Prognostics and Health Systems Management to Predictive Maintenance 2: Knowledge, Traceability and Decision (Reliability of Multiphysical Systems Set – Volume 7)

EL HAMI Abdelkhalak, RADI Bouchaib
Dynamics of Large Structures and Inverse Problems (Mathematical and Mechanical Engineering Set – Volume 5)
Fluid-Structure Interactions and Uncertainties: Ansys and Fluent Tools (Reliability of Multiphysical Systems Set – Volume 6)

KHARMANDA Ghias, EL HAMI Abdelkhalak
Biomechanics: Optimization, Uncertainties and Reliability (Reliability of Multiphysical Systems Set – Volume 5)

LEDOUX Michel, EL HAMI Abdelkhalak
Compressible Flow Propulsion and Digital Approaches in Fluid Mechanics (Mathematical and Mechanical Engineering Set – Volume 4)
Fluid Mechanics: Analytical Methods (Mathematical and Mechanical Engineering Set – Volume 3)

MORI Yvon
Mechanical Vibrations: Applications to Equipment

2016

BOREL Michel, VÉNIZÉLOS Georges
Movement Equations 1: Location, Kinematics and Kinetics (Non-deformable Solid Mechanics Set – Volume 1)

BOYARD Nicolas
Heat Transfer in Polymer Composite Materials

CARDON Alain, ITMI Mhamed
New Autonomous Systems (Reliability of Multiphysical Systems Set – Volume 1)

DAHOO Pierre Richard, POUGNET Philippe, EL HAMI Abdelkhalak
Nanometer-scale Defect Detection Using Polarized Light (Reliability of Multiphysical Systems Set – Volume 2)

DE SAXCÉ Géry, VALLÉE Claude
Galilean Mechanics and Thermodynamics of Continua

DORMIEUX Luc, KONDO Djimédo
Micromechanics of Fracture and Damage (Micromechanics Set – Volume 1)

EL HAMI Abdelkhalak, RADI Bouchaib
Stochastic Dynamics of Structures (Mathematical and Mechanical Engineering Set – Volume 2)

GOURIVEAU Rafael, MEDJAHER Kamal, ZERHOUNI Noureddine
From Prognostics and Health Systems Management to Predictive Maintenance 1: Monitoring and Prognostics (Reliability of Multiphysical Systems Set – Volume 4)

KHARMANDA Ghias, EL HAMI Abdelkhalak
Reliability in Biomechanics
(Reliability of Multiphysical Systems Set –Volume 3)

MOLIMARD Jérôme
Experimental Mechanics of Solids and Structures

RADI Bouchaib, EL HAMI Abdelkhalak
Material Forming Processes: Simulation, Drawing, Hydroforming and Additive Manufacturing
(Mathematical and Mechanical Engineering Set – Volume 1)

2015

KARLIČIĆ Danilo, MURMU Tony, ADHIKARI Sondipon, MCCARTHY Michael
Non-local Structural Mechanics

SAB Karam, LEBÉE Arthur
Homogenization of Heterogeneous Thin and Thick Plates

2014

ATANACKOVIC M. Teodor, PILIPOVIC Stevan, STANKOVIC Bogoljub, ZORICA Dusan
Fractional Calculus with Applications in Mechanics: Vibrations and Diffusion Processes
Fractional Calculus with Applications in Mechanics: Wave Propagation, Impact and Variational Principles

CIBLAC Thierry, MOREL Jean-Claude
Sustainable Masonry: Stability and Behavior of Structures

ILANKO Sinniah, MONTERRUBIO Luis E., MOCHIDA Yusuke
The Rayleigh–Ritz Method for Structural Analysis

LALANNE Christian
Mechanical Vibration and Shock Analysis – 5-volume series – 3rd edition
Sinusoidal Vibration – Volume 1
Mechanical Shock – Volume 2
Random Vibration – Volume 3
Fatigue Damage – Volume 4
Specification Development – Volume 5

LEMAIRE Maurice
Uncertainty and Mechanics

2013

ADHIKARI Sondipon
Structural Dynamic Analysis with Generalized Damping Models: Analysis

ADHIKARI Sondipon
Structural Dynamic Analysis with Generalized Damping Models: Identification

BAILLY Patrice
Materials and Structures under Shock and Impact

BASTIEN Jérôme, BERNARDIN Frédéric, LAMARQUE Claude-Henri
Non-smooth Deterministic or Stochastic Discrete Dynamical Systems: Applications to Models with Friction or Impact

EL HAMI Abdelkhalak, RADI Bouchaib
Uncertainty and Optimization in Structural Mechanics

KIRILLOV Oleg N., PELINOVSKY Dmitry E.
Nonlinear Physical Systems: Spectral Analysis, Stability and Bifurcations

LUONGO Angelo, ZULLI Daniele
Mathematical Models of Beams and Cables

SALENÇON Jean
Yield Design

2012

DAVIM J. Paulo
Mechanical Engineering Education

DUPEUX Michel, BRACCINI Muriel
Mechanics of Solid Interfaces

ELISHAKOFF Isaac *et al.*
Carbon Nanotubes and Nanosensors: Vibration, Buckling and Ballistic Impact

GRÉDIAC Michel, HILD François
Full-Field Measurements and Identification in Solid Mechanics

GROUS Ammar
Fracture Mechanics – 3-volume series
Analysis of Reliability and Quality Control – Volume 1
Applied Reliability – Volume 2
Applied Quality Control – Volume 3

RECHO Naman
Fracture Mechanics and Crack Growth

2011

KRYSINSKI Tomasz, MALBURET François
Mechanical Instability

SOUSTELLE Michel
An Introduction to Chemical Kinetics

2010

BREITKOPF Piotr, FILOMENO COELHO Rajan
Multidisciplinary Design Optimization in Computational Mechanics

DAVIM J. Paulo
Biotribolgy

PAULTRE Patrick
Dynamics of Structures

SOUSTELLE Michel
Handbook of Heterogenous Kinetics

2009

BERLIOZ Alain, TROMPETTE Philippe
Solid Mechanics using the Finite Element Method

LEMAIRE Maurice
Structural Reliability

2007

GIRARD Alain, ROY Nicolas
Structural Dynamics in Industry

GUINEBRETIÈRE René
X-ray Diffraction by Polycrystalline Materials

KRYSINSKI Tomasz, MALBURET François
Mechanical Vibrations

KUNDU Tribikram
Advanced Ultrasonic Methods for Material and Structure Inspection

SIH George C. *et al.*
Particle and Continuum Aspects of Mesomechanics